LIFE EVERLASTING

Books by Bernd Heinrich

Life Everlasting

The Nesting Season

Summer World

The Snoring Bird

The Geese of Beaver Bog

Why We Run (previously titled *Racing the Antelope*)

Winter World

Mind of the Raven

Bumblebee Economics

The Trees in My Forest

One Man's Owl

Ravens in Winter

A Year in the Maine Woods

The Hot-Blooded Insects

In a Patch of Fireweed

The Thermal Warriors

LIFE EVERLASTING

The Animal Way of Death

BERND HEINRICH

MARINER BOOKS

HOUGHTON MIFFLIN HARCOURT

Boston   New York

First Mariner Books edition 2013
Text and illustrations copyright © 2012 by Bernd Heinrich

For information about permission to reproduce selections from this book, write to
trade.permissions@hmhco.com or to Permissions, Houghton Mifflin Harcourt
Publishing Company, 3 Park Avenue, 19th Floor, New York, New York 10016.

www.hmhco.com

Library of Congress Cataloging-in-Publication Data
Heinrich, Bernd, date
Life everlasting: the animal way of death / Bernd Heinrich.
p. cm.
ISBN 978-0-547-75266-2 (hardback)   ISBN 978-0-544-00226-5 (pbk.)
1. Animal ecology. 2. Animal life cycles. 3. Animals — Psychological aspects.
4. Animal behavior. 5. Animal communication. I. Title.
QH541.13.H45 2012
591.7 — dc23
2012010583

Book design by Lisa Diercks
Text set in Miller

Printed in the United States of America
22 23 24 25 26 LSB 11 10 9 8 7

CONTENTS

V. CHANGES

INTRODUCTION

*If you would know the secret of death you must
seek it in the heart of life.*

— Kahlil Gibran, *The Prophet*

*. . . . Earth's the right place for love;
I don't know where it's likely to go better.*

— Robert Frost, "Birches"

Yo, Bernd —

I've been diagnosed with a severe illness and am trying to get my
final disposition arranged in case I drop sooner than I hoped.
I want a green burial — not any burial at all — because human
burial is today an alien approach to death.

Like any good ecologist, I regard death as changing into
other kinds of life. Death is, among other things, also a wild
celebration of renewal, with our substance hosting the party.
In the wild, animals lie where they die, thus placing them into
the scavenger loop. The upshot is that the highly concentrated
animal nutrients get spread over the land, by the exodus of flies,
beetles, etc. Burial, on the other hand, seals you in a hole. To de-
prive the natural world of human nutrient, given a population

of 6.5 billion, is to starve the Earth, which is the consequence of casket burial, an internment. Cremation is not an option, given the buildup of greenhouse gases, and considering the amount of fuel it takes for the three-hour process of burning a body. Anyhow, the upshot is, one of the options is burial on private property. You can probably guess what's coming . . . What are your thoughts on having an old friend as a permanent resident at the camp? I feel great at the moment, never better in my life in fact. But it's always later than you think.

This letter from a friend and colleague compelled me toward a subject I have long found fascinating: the web of life and death and our relationship to it. At the same time, the letter made me think about our human role in the scheme of nature on both the global and the local level. The "camp" referred to is on forest land I own in the mountains of western Maine. My friend had visited me there some years earlier to write an article on my research, which was then mostly with insects, especially bumblebees but also caterpillars, moths, butterflies, and in the last three decades, ravens. I think it was my studies of ravens, sometimes referred to as the "northern vultures," that may have motivated him to write me. The ravens around my camp scavenged and recycled hundreds of animal carcasses that friends, colleagues, and I provided for them there.

My friend knows we share a vision of our mortal remains continuing "on the wing." We like to imagine our afterlives riding through the skies on the wings of birds such as ravens and vultures, who are some of the more charismatic of nature's undertakers. The dead animals they disassemble and spread around are then reconstituted into all sorts of other amazing life throughout the ecosystem. This physical reality of nature is for both of us not only a romantic ideal but also a real link to a place that has personal meaning. Ecologically speaking, this vision also involves plants, which makes our human role in nature global as well.

The science of ecology/biology links us to the web of life. We are a literal part of the creation, not some afterthought — a revelation no less powerful than the Ten Commandments thrust upon Moses. According to strict biblical interpretations, we are "dust [that shall] return to the earth as it was: and the spirit shall return unto God who gave it" (Ecclesiastes 12:7); "thou return unto the ground; for out of it thou wast taken; for dust thou art and unto dust shalt thou return" (Genesis 3:19).

The ancient Hebrews were not ecologists, however. If the famous lines from Genesis and Ecclesiastes had been stated with scientific precision, they would not have been understood for two thousand years; not one reader would have been ready for the concept. "Dust" was a metaphor for matter, earth, or soil. But in our minds the word "dust" suggests mere dirt. We came from and return to just dirt. No wonder early Christians belittled our physical existence and sought separation from it.

But in fact we do not come from dust, nor do we return to dust. We come from life, and we are the conduit into other life. We come from and return to incomparably amazing plants and animals. Even while we are alive, our wastes are recycled directly into beetles, grass, and trees, which are recycled further into bees and butterflies and on to flycatchers, finches, and hawks, and back into grass and on into deer, cows, goats, and us.

I do not claim originality in examining the key role of the specialized undertakers that ease all organisms to their resurrection into others' lives. I do believe, however, that many readers are willing to examine taboos and to bring this topic into the open as something relevant to our own species. Our role as hominids evolving from largely herbivorous animals to hunting and scavenging carnivores is especially relevant to this topic; our imprint has changed the world.

The truism that life comes from other life and that individual death is a necessity for continuing life hides or detracts from the ways in which these transformations happen. The devil, as they say, is in the details.

Recycling is perhaps most visible — as well as dramatic and spectacular — in large animals, but far more of it occurs in plants, where the most biomass is concentrated. Plants get their nutrients from the soil and the air in the form of chemicals — all bodies are built of carbons linked together, later to be disassembled and released as carbon dioxide — but nevertheless they are still "living off" other life. The carbon dioxide that plants take up to build their bodies is made available through the agency of bacteria and fungi and is sucked up massively and imperceptibly from the enormous pool of past and present life. The carbon building blocks that make a daisy or a tree come from millions of sources: a decaying elephant in Africa a week ago, an extinct cycad of the Carboniferous age, an Arctic poppy returning to the earth a month ago. Even if those molecules were released into the air the previous day, they came from plants and animals that lived millions of years ago. All of life is linked through a physical exchange on the cellular level. The net effect of this exchange created the atmosphere as we know it and also affects our climate now.

Carbon dioxide, as well as oxygen, nitrogen, and the other molecular building blocks of life, are exchanged freely from one to all and all to one daily on a global scale, wafted and stirred throughout the atmosphere by the trade winds, by hurricanes and breezes. Molecules that have long been sequestered in soil may be exchanged within the local community over a long time. Plants are made from building blocks derived from centipedes, gorgeous moths and butterflies, birds and mice, and many other mammals, including humans. The "ingestion" of carbon by plants is really a kind of microscopic scavenging that happens after intermediaries have disassembled other organisms into their molecular parts. The process differs in method from that of a raven eating a deer or a salmon, whose meat is then spread through the forest in large and not yet fully disassembled packets of nitrogen, but it does not differ in concept.

DNA, on the other hand, though made mainly of carbon and nitrogen, is precisely organized and passed on directly from one individual plant or animal to the next through a fabulous copying mechanism that has operated since the dawn of life. Organisms inherit specific DNA molecules, which are copied and passed from one individual to another, and so it has continued over billions of years of ever-conservative descent, which has branched through innovation into trees, birds-of-paradise, elephants, mice, and men.

WE THINK OF the animals that do the important work of redistributing the stuff of life as scavengers, and we may admire and appreciate them for providing their necessary "service" as nature's undertakers. We think of them as life-giving links that keep nature's systems humming along smoothly. We tend to distinguish scavengers from predators, who provide the same service, but by killing, which we associate with destruction. But as I began to think about nature's undertakers, the distinction between predators and scavengers became blurred and almost arbitrary in my mind. A "pure" scavenger lives on only dead organisms, and a pure predator on only what it kills. But very few animals are strictly one or the other. Ravens and magpies may be pure scavengers in the winter, but in the fall they are herbivores eating berries, and in the summer they are predators living on insects and mice and anything else they can kill. Certain specialists, however, some with unique abilities, spend most of their time finding food in one way. Polar bears usually catch seals at their breathing holes in the ice, but on occasion they will find and eat a dead one. A grizzly bear will relish a dead caribou as well as one it has killed, but most of the time it grazes on plants. A peregrine falcon is a swift flyer that captures flying prey, while a vulture would not as a rule be able to capture an uninjured live bird, so it has to rely on large, already dead prey. Indeed, vultures, ravens, lions, and almost all of the animals we typically typecast as "predators" just as readily take the ailing and half-dead and the (preferably fresh) dead; they will not enter a fight for life with another animal unless they have to. Herbivores too take those organisms that are least able to defend themselves. Deer and squirrels, for instance, munch on clover and nuts but will gladly eat any baby birds that they find in a nest. Strictly speaking, herbivores take the most lives; an elephant kills many bushes every day, while a python may ingest but one wart hog a year.

The potential ramifications of recycling are almost as varied as the number of species. I hope to provide a wide view, and I give examples from personal experiences everywhere from my camp in Maine to the African bush.

I

SMALL TO LARGE

Size is an important aspect of the way an organism can live and the form it can have. It determines the kind and proportions of body support systems that are needed to fight gravity. An organism's size determines the diffusion rates of gases and nutrients, which set the maximum metabolic rate, the amount of food required, the kinds of places used to hide in, and the defenses needed. Size matters in how a body is disposed of, who the disposers are, and how they do it. Burial, part of what we associate with "undertaking," is seldom part of disposal, but when it is, it is not to get rid of bodies but to keep them for a purpose.

BEETLES THAT BURY MICE

I often see flowers from a passing car
That are gone before I can tell what they are.

— Robert Frost, "A Passing Glimpse"

CATS MAY SCRAPE LEAVES AND GRASS OVER DEAD PREY TO conceal it, and some wasps drag drugged but living insects into previously constructed homes so the wasp larvae can safely feed on fresh meat. But to my knowledge, only one group of animals, beetles belonging to the genus *Nicrophorus,* regularly moves carcasses to a suitable place and then deliberately buries them. Unlike humans, who generally bury only our own species and those pets who have become surrogate humans, these beetles bury a great diversity of birds and mammals but never their own kind. They bury dead animals as a food source for their larvae, and the burying is a central part of their mating and reproductive strategy.

There is much in a name, but sometimes it can be misleading. The genus name, *Nicrophorus,* comes from the Greek *nekros,* meaning dead, and *philos* or *philia,* meaning loving, or perhaps *phoms,* meaning to carry. (*Nicro* was probably a spelling mistake by the person who originally named the species; by scientific convention, the original name has precedence.) Not that "death-loving" is strictly accurate, either. It might factually have been more appropriate to call the beetles "life-loving," or *viviphorous,* because their whole purpose in seeking out dead animals is to create life from those already dead. A mouse carcass, for example, will nourish a dozen or more new beetles.

Burying beetles (also called sexton beetles) are master mouse undertakers. They are strikingly beautiful, colored deep black and adorned with flashy, bright orange markings on the back. Their fascinating life cycle involves monogamy and extensive parental care of their offspring. They are so common and widespread that almost anyone in the north temperate zone of the earth can see them in late summer if they want. I meet them regularly every summer, but only because I give them dead mice and road-killed birds.

The romantic story line of *Nicrophorus* beetles (of which there are sixty-eight recognized species worldwide, including ten in northeastern North America, where I live) is that a pair is formed after meeting at a carcass. When a male finds a mouse or other suitable carcass, he executes what is essentially a headstand on it and emits scent from a gland at his rear end. The "calling" scent wafts away on a breeze, and if a female detects it she flies upwind to find him and his carcass, and they mate. (On the other hand, if a male comes along, he may be aggressively excluded by the original mouse claimant.) The male and female cooperate in burying the carcass in part to get it away from other possible claimants, which often requires transporting it to soil suitable for digging the burial chamber.

Burying beetles don't have grasping feet, so the pair transport a carcass by crawling underneath it and lying on their backs with feet up, "walking" on the carcass rather than on the ground. As long as the beetles can partially lift the mouse while their backs stay pressed to the ground, the carcass moves forward. The trick is to get it to move in the desired direction, and indeed these beetles do pick a specific direction and stick to it. The wonder, I think, is that both members of the pair apparently "know" where they are going, because they work the carcass in the same direction rather than at random, which would not be at all effective.

After the beetles have moved the carcass to the selected area, they excavate the soil under the carcass by pushing it out to the sides. Gradually a pit develops and the mouse (or other small animal) carcass, which is by now softening, folds inward and gradually sinks into the soil. After burying it several inches deep, the pair continue to roll the carcass into a ball while at the same time removing its hair (or feathers). They spray it with antibiotic secretions from the anus, which kill bacteria and fungi and thus retard spoilage of this valuable food. The female then lays eggs in soil nearby. The larvae hatch in several days, crawl to the carcass, and settle into a depression at the top. Both parents feed the larvae regurgitated food taken from the carcass until the larvae are able to burrow through the skin and into the softening meat.

Like the parents of altricial baby birds (those that hatch naked and helpless), the adult beetles make squeaking noises to announce that they're ready to feed the young. In response, like baby birds, the grubs raise themselves up and are fed directly, mouth to mouth. As the grubs get older, they feed themselves, but they may still be assembled by the parental dinner call and fed directly. After several days the male usually comes up out of the ground to search for another carcass and start another family. The female usually stays longer with the larvae.

After a week to ten days, the fully grown young burrow into the surrounding soil, where they pupate. In most northern species, they hibernate where they are and emerge as adults the following spring or summer, with the seasonal timing varying among species.

The *Nicrophorus* life cycle has been studied in ever greater depth and detail for more than a century and is still a source of wonder. Current work has centered on hormonal regulation of the life cycle and on differences between species of burying beetles. For example, one species buries snake eggs rather than animal carcasses. Usually the behaviors follow the pattern described above, as notes from my journal suggest:

11 August, 2009. 5 P.M. The fresh mouse [white-footed, *Peromyscus leucopus*] I put out this morning is no longer visible; it is almost buried by a *Nicrophorus*. Only the one beetle on it when I pulled the carcass out of the ground.

12 August, 2009. The mouse is re-buried, when I checked on it at 3 P.M. As expected, there is now a pair on it.

That year I returned to my camp on August 27, and ten days later I excavated the buried mouse; I found only a skull and a clump of fur, along with fifteen to twenty *Nicrophorus* beetle grubs stacked up vertically in tight formation, consuming the last of the mouse. There were no fly maggots.

On my previous visit ten days earlier, I had set traps in the cabin to catch more mice, and I now had five dead ones in various stages of decay, from soft and stinking to dried up (and also stinking). Before setting them on the ground outside for the beetles, I tied a white string onto each carcass with a different number of knots at the end of each string so that I could identify them after they were buried. Only a few minutes after I deposited the mice, the first sextons came humming along and homing in on the scent. Seconds after they flew in, they plopped onto the ground and, with their antennae waving, made a beeline to a mouse.

In two hours one mouse had seven beetles crawling over and

under it and squeaking. I was surprised to see so many beetles at one mouse, but this one was half dried up and could not serve as a brood nest, so it was not likely being strongly defended by a pair wanting to claim it. I was also amazed at the noise-making, because beetles don't have ears; the sound is made by the friction of body parts rubbing together. Shiny greenbottle flies (botflies) came, ready to deposit eggs that would in hours hatch into hungry maggots. Occasionally, white-faced hornets on the hunt for flies flew in and patrolled close to the ground, pouncing down here and there, but their potential prey usually disappeared shortly after the hornets arrived.

Beetle pairs met at two other mouse carcasses and within an hour had buried them by excavating the earth to create a hole into which the mouse dropped. Thus was their prize removed from the competitors: ravens, botfly maggots, and other beetles. The beetles had mites all over their bodies, as though infested with parasites. But these mites are the beetles' allies, for they kill or consume the botfly eggs that got onto the mice before they were buried.

The next summer I offered up a shrew, a fresh *Blarina brevicauda*, in order to again watch the undertakers. This shrew is one of the few mammals with a poisonous bite and an unpleasant odor, and even most predators who kill one then discard it. House cats commonly bring them indoors, and most people think of them as moles because they have short gray hair and a pointed nose, though not the shovel-like front feet of a mole. Shrews are among the most common animals in the north woods, but those of this species are seldom seen because, unlike most shrews, they live underground.

It was August 5, 2010. I had brought the *Blarina* shrew to camp late the night before and put it outside by the camp door in a clean spaghetti-sauce jar tipped sideways. The next morning, at 6 A.M., after I had had my coffee and toast and was ready to settle down for a long uninterrupted session of beetle watching, I went out for a look. Four beetles were in attendance, and in their gorgeous jet-black garb with bright orange stripes they made a pleasing picture on the shrew's dark gray fur. They were all *Nicrophorus tomentosus,* a species that has short yellow fuzz on the thorax. They had already removed the shrew from the jar, and a pair of them were under it, heaving it along the ground beyond the jar lip. The other two beetles, who were much smaller, were ten centimeters distant and apparently hiding underneath the jar cap. They stayed there for at least the next hour as the larger pair continued to move the shrew.

The ground near my doorstep is solidly packed; it was not a suitable place for the beetles to bury their prize. Independently of each other, the two beetles made repeated excursions in all directions, sometimes as far as sixty centimeters away, as though searching for a place to take the carcass for burial. Then they would return to the shrew before going off in another direction. How did they find their way back each time? Did they remember the route they had taken? To find out, I intercepted a beetle and placed a spoon in front of it, guiding it to walk onto the spoon, then released the beetle two feet from the carcass. It walked directly to the carcass without any apparent search. Had it memorized the local topography? I intercepted the other beetle as it was searching sixty centimeters north of the carcass and released it a meter and a half south of the shrew. If it navigated by remembering its previous route, it should now continue to run on away from the shrew. This beetle stayed still for a few moments, then lifted off in flight and went directly back to the carcass. I now removed one of the two beetles a meter and a half away, as before, and this one preened and then flew directly back to the carcass as well. Another time I placed one of the beetles a meter away from the carcass. This one turned in many little circles as though searching before walking in a direct line back to the shrew.

These beetles appeared to be more clever than I thought possible. I would have liked to experiment further, and the mystery of their homing ability remains in my mind. But I had to stop interfering with these individuals, because at this time I was mostly concerned with their carcass handling.

Meanwhile, one of the smaller beetles that had hidden under the jar cap emerged and went directly to the carcass. Would it try to steal it away from the pair? No, its intentions were clear; the second it got there it (he!) jumped on the back of the larger of the pair and copulated with her. That took only a few seconds, and then he left immediately and hid again, this time under loose bark on the soil about thirty centimeters away. There was a lot more going on here than I had ever expected, so I kept up my watch.

At 7:15 A.M. another *Nicrophorus* came flying in. This one circled around for at least a minute before it finally landed at the shrew carcass. Almost immediately it mated with the resident female. Then still another flew in, and the same thing happened. The female who was getting all of this attention continued crawling around the shrew without any interruption — around and under the carcass she went, trying to bury it. I thought that the pair was "supposed" to fight the intruders off, but I saw no fighting. By this time the shrew had been moved only about thirty centimeters away from the jar, but it was still on hard-packed ground, and there was no soft soil nearby.

After the second mating, which again took only a few seconds, I picked up the visiting male with my spoon (so as not to handle him directly and startle him) and dropped him a meter and a half from the carcass, wondering if he would return to it or to her. He seemed not in the least perturbed; he just sat where I dropped him and proceeded to clean himself compulsively. He used his legs to rub his abdomen, then his head and antennae, and then he rubbed his legs against each other. Done with that, he hesitated a few moments, then took off in flight and circled around before landing on a cherry twig at least three meters distant from the shrew, where I took photographs of him. I noticed then that he was covered with mites. I had not seen mites at all while he was mating; at that time the bright orange stripes on his back stood out. Now the mites were attached all over his back, almost fully obscuring the orange so that he looked pinkish brown. The mites were camouflaging him. But where had they been just *before* he took flight?

To the mites, the beetles are simply a vehicle for hitchhiking to more fresh fly eggs. As soon as a beetle finds a carcass, the mites hop off and forage, and when the beetle leaves they presumably hop back on.

A friend recently told me he caught several sexton beetles in a can with some rotting meat. When he later looked in he found that two were dead and two almost dead. He saw mites "walking all over the half-dead ones as though to revive them." One of the sextons did revive, and then all the mites on it immediately scooted under its elytra (wing covers). I wondered if the mites sensed that the beetle was leaving. This is not a preposterous idea (although I am of course not inferring conscious knowing) because the beetle would shiver before taking flight to raise its body temperature, and either the vibrations or the temperature change might signal the mites to attach themselves so they could be safely carried.

Another beetle came flying in around 8:00 A.M., apparently with the same intention as the others, and it was not just to eat carrion. First things first: an immediate mating. Still another one came at 8:15. For a while there were five beetles at the carcass, but three eventually walked off and hid under the debris on the ground nearby. The original pair was finally alone at the carcass. I'm not sure what all of these vignettes prove, but they did cast doubt in my mind about the beetles' reputed monogamy.

Throughout the two and a half hours I watched the beetles, I heard repeated episodes of squeaking. These sounds occurred only at the carcass, and I think they were made by the pair, because I heard the sounds as much when they were there alone as when others joined them, so the squeaking probably was some kind of communication between the two.

As the air temperatures rose through the morning, both green and blue botflies started flying in. A few red *Formica* slave-making ants, which had earlier in the summer moved into the roof space of my cabin, also showed up. The beetles appeared to pay no attention to these visitors, but by now it was clear that this shrew was not going to get buried. The distance to suitably soft soil was too great for the pair to move it that far. The prize would now go to the flies or the ants or would become a resource for food or sex for adult burying beetles. It was not destined to be a place and resource for rearing their young.

I watched the beetles try to move the shrew for the rest of the day to document the developing story as fully as I could. By noon, when temperatures in the shade had risen to 86 degrees F, as many as eight beetles — all *N. tomentosus* — were at the shrew carcass at the same time. I had watched nearly continuously and observed fourteen additional matings. On the other hand, I had seen only two tussles, each lasting only seconds — no real fighting. The pair eventually moved the shrew about a meter farther before they stopped trying; it was still on packed soil. Though botflies were abundant, I saw no eggs. I observed one large muscid fly deposit a live larva but saw no developing maggots.

By noontime the beetles had finally chewed a hole through the belly skin of the shrew. At least two entered the carcass and were presumably feeding there or else trying to find a way out, as the skin heaved here and there. From then on the carcass did not budge from the spot. By 3:00 P.M. there were no beetles left, but that night, at 8:45 P.M., I found two back under it.

The next day I opened the carcass and found seemingly fresh meat but not one maggot, despite the dozens of both green and blue botflies that had been on it the day before, along with the beetles. I moved the shrew onto soft soil, and for the rest of that day there were never more than two beetles on the carcass, which the pair finally buried.

MICE AND SHREWS are of manageable size for these undertakers, but what would happen, I wondered, if they got something far too big, such as a super-sized "mouse" — a fresh road-killed gray squirrel? In the first two hours after I laid the cut-open squirrel on the ground next to my cabin, it attracted five *N. tomentosus*. The next day there were up to eighteen of these beetles on the carcass at a time, with one to four of them "calling" (standing still with hind end up in the air dispensing their come-hither scent to attract others). More beetles were flying in while others were leaving. I saw no evidence of pairs, though; most of the beetles were feeding on the meat, and there was a lot of random-seeming mating.

The next day the temperature dropped from 75 degrees F to 55. There had been very few botflies, and now all of the beetles left. Most walked from a half meter to two meters away from the squirrel carcass and buried themselves under the leaves and in the soil. A raven came and took away the carcass with its still almost fresh meat, putting an end to my attempt to find out what the beetles might do.

I decided to let a rooster stand in for the squirrel so I could get a more detailed picture. I laid a fully feathered dead bantam rooster belly-down (but not cut open) on the ground in the woods. Temperatures were in the high 80s, and when I checked the rooster the next day, it had already attracted hundreds of green botflies, whose maggots could devour the whole animal in days. The rooster's feathers were coated with hundreds if not thousands of white botfly eggs. When I turned the bird on its back, more than a dozen *Nicrophorus* beetles scattered in all directions. The dry leaves surrounding the rooster fizzed like an opened champagne bottle. All the sextons were running fast and burying themselves under the loose leaves. I was so startled to see them I didn't catch a single one.

At this carcass I noted several species of *Nicrophorus*. I had not known how much variation in coloration and size there is between species, and I wondered if I should examine all of the beetles in detail to find out what species they were. Should I wait to see if they would fight it out until only one pair of each species remained? Or would all the beetles somehow "cooperate" and fight off the imminent takeover by the flies? I decided to just wait and see what would happen.

By the next day there were even more *Nicrophorus* beetles on the rooster carcass. Curiously, there were no more botflies, although temperatures continued to hold in the 80s. Even more curious, all of the botfly eggs that had been deposited on the feathers now seemed shriveled; they had apparently died. There was not a maggot in sight. The skin of the rooster looked pristine, as though it had been sterilized. I was seeing the result of the sexton beetles' battle with the flies, and here the beetles had won. The numerous (about two dozen) beetles at this large carcass had probably reduced the competition from maggots, which usually overwhelm a carcass and claim all or most of it. With so many beetles, though, each one paid a price, because when two or more pairs are breeding communally at a carcass, the larger individuals are stimulated to increase their ovarian development, while the smaller beetles' development is delayed. I was glad that I had let the drama play out, yielding such a clear and dramatic result.

To determine how many beetle species were using the carcass,

I needed to catch them all. I left the beetle horde to reassemble under the rooster for a couple of hours, and when I came back I carefully removed all of the leaf litter and loose soil around the carcass to limit the beetles' escape routes. Then, keeping an open jar at the ready to dump the beetles in, I turned the rooster over and started grabbing. After I caught all the running beetles I could see, I started digging. Those that had made it under cover were not always easy to see, because the second defense of these beetles who bury the dead is to fake death; they curl up with their legs extended, just like a dead specimen. They lie on their side or back so that you see only their dark underside against dark soil; the bright orange patches on their backs are not visible. Despite their evasive maneuvers, I managed to get a haul of thirty-nine *Nicrophorus* beetles, which turned out to be of four different species.

I followed the fate of the rooster carcass for another five days, continuing to collect beetles, for a total of seventy sextons (fifty-eight *N. tomentosus*, nine *N. orbicollis*, two *N. defodiens*, and one *N. sayi*). The *N. tomentosus*, by far the majority, were the same species I had observed in pairs on the many mouse, shrew, and chipmunk carcasses I had set out that month. There had been no burial of this rooster, but neither had the flies taken it; it remained maggot-free. On the sixth night it was finally taken away by a large animal, probably a skunk or a raccoon.

THERE WERE MORE mysteries here than I could ever solve, making these results all the more exciting. But as always, some of the most surprising revelations from my observations had nothing to do with what I was originally looking for. In this case something unusual caught my eye when I dumped some of the captured beetles into a second jar.

As I have mentioned, the first escape strategy by these beetles when a "predator" comes to the carcass they are on is to run and hide and then burrow into the soil. I've described how they "play possum," feigning death, but when I picked them up they quickly gave up on that defense and bit me instead. In the jar, with these three options unavailable, some beetles tried a different tactic: flying out. I looked closely at a beetle in my jar, admiring its black back with bright orange markings — not only because the markings were beautiful but also because they can identify the species. I was shocked to see the bright orange and black turn, in a flash before my eyes, to brilliant lemon yellow the instant the beetle took flight! How could that be?

Like many insects living today, the ancestors of beetles had two pairs of wings. Now, however, they have only one pair. The original first pair has become modified into a two-part hard shell, the elytra, or wing covers, which are useless in flight but serve as a coat of armor covering the wings when they are not in use. The elytra are often decorative. The beetle's membranous wings are usually at least twice as long as the animal's body, but when not in use they are stored under the elytra, folded up like sheets put away in a drawer. In most kinds of beetles, the wing covers are usually spread passively to the sides during flight or simply folded over the back. In either case, an observer sees no color change. But I *had* seen a dramatic color change. Or was I seeing things?

I realized I had never seen any orange on the beetles flying near carcasses. I had seen yellow, which I had assumed was because the thorax of *N. tomentosus* is covered with yellow fuzz. Now I wondered if I had overlooked the orange and black because of the beetles' fast, erratic flight. I looked again: yellow only. I decided the camera might see what I could not, so I took some wild shots of flying beetles. I managed to get several fuzzy pictures, which were enough to confirm my hunch: the backs of the beetles in flight were yellow!

I then examined both live and dead beetles, trying to flip their elytra to the sides, as they would be in flight, to expose the top of the abdomen, which I found was black. But as I tried to lift the elytra of both living and dead beetles, they spontaneously rotated, twisting up with the outer edges turning inward. Then, when I moved the elytra back, they locked into position to cover the abdomen so that the formerly outer surface was now inward. In other words, the elytra, unlike those in any other beetles, as far as we know, lock over the back with the *dorsal side down and the previously hidden ventral side up*. And this previously hidden ventral side of the wing covers is . . . lemon yellow! Thus the "secret" of the beetles' color change is that this yellow *underside* is exposed when the beetle flies. In all other beetles in the world (to my knowledge), the upper elytral surfaces remain up during both rest and flight.

This mechanism for an instant color change, although perhaps known, had not been described in the scientific literature. But what was its significance? What could it be for? The color-change mechanism was unique to this species, and because it hid the bright orange, it helped make *N. tomentosus* a convincing mimic of a bumblebee in flight, which suggested an interesting function.

Most of the forty-six or so species of North American bumblebees have black bodies marked with yellow pile. Seven of these species also have varying amounts of orange, but that color is always bordered by yellow and is never in sharp contrast to a black band, as in the burying beetles. When *N. tomentosus* are flying in late summer, there are usually numerous worker bees of up to seven black-and-yellow species (*Bombus affinis, B. vagans, B. bimaculatus, B. sandersoni, B. impatiens, B. perplexus,* and *B. griseocollis*) that have nearly identical color patterns and are difficult to differentiate. With its elytral flip, the airborne *N. tomentosus* becomes, in an instant, a credible mimic of any or all of

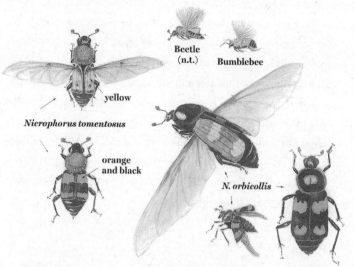

Beetle (n.t.)

Bumblebee

yellow

Nicrophorus tomentosus

orange and black

N. orbicollis →

Note: pictures not to scale (except top beetle vs. bee)

Two of the common species of burying beetles that use small animal carcasses, showing Nicrophorus tomentosus *(left) and* N. orbicollis *(right) in flight and on the ground. In both species the wing covers (elytra) are twisted, unlike those in other beetles: the undersides of the wings become dorsal and hide the bright upper sides, so that the animal in flight mimics the common yellow bumblebees.*

these bumblebees. Few birds will tackle a bumblebee because of the risk of being stung in the mouth. Unlike most other sextons, *N. tomentosus* can thus fly in the daytime and search for carcasses that other sextons must hunt for in dusk or under cover of darkness.

Most burying beetles can appear to mimic bumblebees simply because they are in the same size range and therefore hum like them in flight. But *N. tomentosus* has gone a giant step further by acquiring yellow fuzz on its thorax and by evolving yellow color on the underside of its wings (in museum specimens the colors fade).

I examined the colors of the wing undersides in fresh specimens of the other three species that I captured at the rooster carcass. None of them were lemon yellow, although *defodiens* was orange-yellow; *orbicollis* and *sayi* were grayish or dirty white.

This wing-cover flip mechanism probably also explains the anomaly of the mites I had seen on the back of the beetle *the moment* it landed. While the beetle was in flight, they had been attached not only on the thorax but also "under" the elytra, a logically safe spot to be, and had not yet changed position at the moment the beetle landed and reversed its wing covers to show once again the orange markings on black.

Details of the wing-cover twist of an N. tomentosus *before taking off in flight and on landing, and (center) one of several bumblebee species that occur concurrently with the beetles. The sequence shows the color change from orange and black (three lower beetles) to yellow (three top beetles) in flight.*

THE IDEA OF watching burying beetles as an interesting activity was probably passed down to me by my father. And about twenty years ago I shared this interest with my older son, Stuart, then ten, who was staying with me at camp. I remembered his excitement, which I had written about in *A Year in the Maine Woods*. So I consulted that book (page 256) to find that we had put a dead deer mouse on some sawdust in back of the cabin "to see if a beetle would bury it." When we checked an hour later, the mouse was gone, but Stuart found where it was buried and dug it up. I then put it at a new place, and this time Stuart sat down to watch. What he saw and said surprises me now. He saw one beetle alternately digging and staying stationary while elevating its abdomen (dispersing scent to attract a mate). Then, seeing another beetle fly in, he declared that "it sounded just like a bumblebee and I saw it with its wings still open just after it landed. It had gold fuzz on its back just like a bumblebee." I now know that the "back" he refers to was the yellow of the overturned elytra covering the black abdomen. At the time I probably thought he was referring to the beetle's head and thorax only. Seeing with a child's eyes is seeing without preconceptions, and it is also, when associated with knowledge, the precondition for making discoveries. My discovery about the instant color change involving the rotation of the elytra was worthy of a report in a scientific journal, so I wrote up my findings and submitted the article to *Northeastern Naturalist,* where it was reviewed by scientific colleagues and accepted for publication.

BURYING BEETLES ARE still common, as anyone can discover by putting out fresh meat. You don't have to go looking for them; they will come to you. Nicrophorids are for the most part not endangered. Nevertheless, *Nicrophorus americanus,* the largest beetle of the group, averaging about three centimeters long but sometimes reaching four, is on the list of endangered species in the United States. It has disappeared from 90 percent of its former home range, which included at least thirty-five states, and is now found in only five. Unlike all the other sextons, most of whom are half as long, *N. americanus* is splashed with bright orange-red on the head, thorax, and antennae. Not enough is known about its biology to explain why it is so large, has so much orange-red, and is endangered while most other species are not. Nicrophorids have some peculiar specialties of habitat and food. One species, *N. vespilloides*, buries its carcasses only in peat moss; another, as mentioned earlier, buries snake eggs rather than mice or other small carcasses. One hypothesis is that *N. americanus* specialized on passenger pigeons and that now there are not enough carcasses of similar size regularly available over most of its range.

SENDOFF FOR A DEER

It must be I want life to go on living.
— Robert Frost, "The Census-Taker"

I HAD BROUGHT ANOTHER ROAD-KILLED GRAY SQUIRREL with me to camp. The weather in Maine in mid-June 2011 was too cool for botflies to be active, so the squirrel carcass had no maggots. But it was not too cold to rot — I was pretty sure the squirrel smelled good and ripe. If it did, would a vulture find it? And what would it do with the smelly carcass? Would it swallow it whole, as a great horned owl probably would? To find out, I left the bloated squirrel in a clearing in the woods, then made myself comfortable on the couch next to the window in my cabin.

A raven flying over the forest saw the carcass first. It swerved to the clearing and then perched silently on the top of a pine tree. After surveying the area for a few minutes, it spread its wings and swooped down to the ground beside the squirrel, hopped up and down a few times, then pulled out the eyes and some fur. But it was unable to tear through the skin. Instead it entered the squirrel through the mouth and pulled out a little meat and the brains.

Then it flew off. I ran outside, slit the carcass open, and again made myself comfortable on the couch, hoping to see the raven return.

With the rotting entrails exposed, there would be a strong odor plume. Sure enough, less than an hour later I saw a shadow pass over the ground; a large bird was flying overhead and circling the clearing — a turkey vulture. A minute later it swooped directly over the carcass and, after circling for another minute, landed on the lone apple tree next to the squirrel. There it constantly turned its head, seeming to look in all directions except at the squirrel. After a while it preened, then spread its wings out to the sun, held them stationary, and nonchalantly preened some more.

Watching the vulture through binoculars from inside the cabin, I thought it looked gorgeous. Not a speck of dirt was visible on it anywhere. Its long, ivory-colored bill glistened. The vulture was showing its heightened emotion by blushing; its naked head had turned cherry red from the blood shunted there. Its upper neck, though speckled with a few sparse black feathers, was soon bright purple. Below the almost-bare neck skin shone a thick, shiny, black-blue ruff, contrasting with the dull brown wing feathers.

After sixteen minutes of perching on the apple tree, the vulture started to shift its attention from the surroundings overtly to the squirrel. Hopping closer from branch to branch, the vulture finally dropped down to the ground next to the squirrel. It stood stock-still for several minutes before starting to take delicate, skimpy bites. It pulled on entrails and flung them to the side, then resumed tearing and taking tiny bits of meat. After thirty-four minutes it flew off, leaving the intestines and the almost-clean bones and skin. I left them, wondering if something else might come along to take these leavings.

I woke at sunup the next morning to the sound of a crow, and as I looked up from my bed by a window I saw it perched on the tiptop of a tall spruce near the apple tree. The spruce was swaying in the breeze, and at times the twig on which the crow was standing bent over from the bird's weight. But the crow kept its balance, cawing at the same time, for at least ten minutes. It was facing the squirrel remains, less than a hundred meters away, and I expected it to fly down at any minute. But it just kept up a vigorous cawing.

After a while a second crow chimed in from the valley below; when it arrived, the two flew to the apple tree where the vulture had perched the day before. Finally one of the pair flew down to the squirrel, landed near it, looked at it for about a minute, then flew back up into the apple tree. Both crows stayed in the nearby trees for a few more minutes and then, just as the sun was coming up, they flew off down the valley in silence.

An hour later a raven came flying in, and without a moment's hesitation fluttered directly down to the squirrel carcass. It picked up the remains in its bill and flew off into the dense lower branches of the spruce. The raven was almost hidden there, but I saw it begin to pluck out fur. It quickly stopped and left the skin on a branch. That afternoon it returned, flying directly to where it had left the squirrel skin. There could not have been much left; when the raven flew off I found the skin, turned inside out (so the fur was inside), on the ground. A day later that too was gone. A larger carcass, I hoped, might bring even more of a crowd.

July of 2010 was hot and muggy in Maine. On the ninth, my friend Wallis, a builder in the village of Weld, near where I stay, and I were sweating as we put up the framework of a sauna. The heat, though oppressive for us, was ideal for deer flies. Ten to twenty were circling us at any given time, each one looking for an opening to press its attack—to make a cut in the skin to lap up fresh blood. I do not appreciate those who feed from me before I am dead, and every one of these flies intend to do just that.

They scope you out by noisily zooming around you and then, in an unguarded moment, they land on any available bare spot of skin. A single deer fly doesn't bother me, because I have learned their tactics and I know how to kill them. But with twenty at a time the odds are stacked in their favor.

These flies don't bother just humans; they also drive the moose and deer in the woods to distraction. Moose may escape by submerging in pond water. Deer, which don't browse on pond weed, have only the option of running and leaping. Apparently they don't always look before they leap. That day, while Wallis and I were driving between Dixfield and Weld to get the cedar planking, we found a dead doe that had gotten smacked, and not too recently; it already smelled pretty ripe. I can't imagine driving along, hitting a deer, and then leaving it by or on the road. And no one else had stopped either.

I did stop, as I often do, and I was sorry to find that the doe's teats showed she had been lactating; somewhere nearby in the woods one or maybe two fawns would be waiting for milk and soon starving to death. It seemed a shame to let the deer carcass go to waste. With Wallis pushing and me pulling on a hind leg from the back of his pickup truck, we got her aboard and drove off. I wasn't yet sure what to do with her, but I hoped that coyotes, bear, ravens, or vultures might find her in the clearing by my cabin. They would then feed their young and convert her into other lives, and in the short term they wouldn't have to catch and kill something else.

I dropped the doe off among the goldenrod and meadowsweet bushes in my clearing. With my sharp hunting knife I opened her belly to spill the entrails, laying the scapula and right front leg over to the side. I also sliced the skin to expose the meat of a hind leg. Then, taking a break from building the sauna, I took up my post on the couch by the cabin window to see what would happen.

Two hours later the first turkey vulture came soaring in and started circling over the doe. A couple of times before I had seen swarms of vultures hunkering in the trees near a deer alongside the highway, perhaps waiting for traffic to stop. But my pair of ravens nest in the nearby pines, and they now had large, hungry young. This is going to be good, I thought. Here in the field, this could be a circus.

The previous fall my nephew Charlie, who lives in suburban Pennsburg, Pennsylvania, had claimed a road-killed deer in front of his house. He told me that "within an hour" after he gutted it there in his front yard, several turkey vultures arrived, and then more than a dozen gathered at the very fresh meat. They perched on the roof of his house for a short while and then, after apparently reaching some consensus, they descended together. By the next day not a scrap of meat was left. Only the stomach contents — partially digested corn the deer had harvested from neighboring fields — remained. I wondered how long it would take the vulture crew to arrive here at my deer, and if they would have to fight off the ravens or vice versa.

The vulture flew close over the carcass, rocking from side to side as these birds typically do. It then flew wider around the clearing and the nearby woods, as though checking everything out. But to my disappointment it apparently didn't like what it saw, because it left without even landing.

From the work of the biologist Patricia Rabenold and from Charlie's vulture experience, I knew that vultures have communal roosts, from which naive birds follow experienced ones to any communal feast. I expected a crowd of turkey vultures to arrive the next morning, so later in the afternoon I took the opportunity to do some errands. When I came back about two hours later, a vulture had arrived. It was perching on a tree at the edge of the clearing but flew off as soon as I came. It had not fed on the doe, though; nothing had been disturbed there.

The deer carcass was by now being discovered by hundreds if not thousands of blowflies. There are many blowfly species (1,100 have been described), and most require microscopic examination to differentiate, because in many cases you have to count the number of bristles to distinguish them. The common green blowfly, *Lucilia sericata,* has three bristles on the dorsal side of the mesothorax and six to eight bristles on its occiput (part of the head). *Lucilia cuprina* has only one occipital bristle. I didn't count the blowflies' bristles. I was already sufficiently impressed by the sheer brilliance of the green flies and their obvious difference from the blue ones, of whom there were fewer but whose metallic coloring was just as spectacular.

The carcass covered by these brilliant flies stank. I knew that a raven, as well as two vultures, had seen the carcass, because it circled the clearing once and called several times before flying on. Ravens like their meat fresh or, if not freshly killed, at least frozen. By dusk there were still no birds at the carcass, but after I went to bed that night I heard a coyote concert from the forest up toward Gammon Ridge. From upstairs I watched the clearing as the Big Dipper slowly rotated to the horizon. I strained to see gray shadows creep in toward the doe. But I saw none and soon slept well.

No ravens, crows, vultures, or coyotes came to the doe. But there was plenty of activity the next morning. Ten of my students from the winter ecology course I had taught at this site the previous January had come in the night and set up two tents in the field. Today was a designated party time, so ravens, vultures, and coyotes would stay away. I didn't mind — here was a chance to find out what would happen to the doe when the big guys didn't get to eat. The strictness of the big birds' exclusion was assured by noon, when a second party of revelers arrived, bringing two bottles of elderberry wine and another guitar.

The gathering inside the camp that afternoon, and then outside around the campfire in the evening, was a fitting wake for a deer. Especially if she could appreciate the maudlin sounds of some ten voices chiming in and out to the accompaniment of two guitars, a banjo, and a mandolin. I might be envious, were it my turn to return to the cycle of life.

The next morning, while almost everyone was still zonked out, the doe carcass still lay untouched by birds or coyotes. At dawn a raven again flew by, but this time it remained silent. By noon the temperature had soared to the low 90s, and a lone vulture showed up, circled, and perched in a nearby tree. It didn't fly down. Was something wrong? I checked the carcass. No meat had been torn off, but it was peppered with thousands of green blowflies.

The doe reeked. I got whiffs of her even at the cabin. The chemical of putrefaction that is so offensive to us, ethanethiol (ethyl mercaptan), is, according to the *Guinness Book of World Records*, "the smelliest substance in existence." For humans anyway. It is added in trace amounts to odorless propane so that we can detect it and not blow up our homes by lighting a match. Turkey vultures, supposedly attracted to even minute amounts of ethanethiol, have been used to detect the locations of leaking gas pipelines. However, the intensity of stink does not equal the degree of their attraction; turkey vultures come more readily to fresh or nearly fresh meat.

As I examined the doe on this second day, I saw that the blowflies had won the competition for her. The flies reputedly can smell a carcass from ten miles away, and they had indeed come in swarms, undeterred by putrefaction. The exposed meat was baking black in the sun, and the fur was covered with white patches — masses upon masses of botfly eggs.

More than a half-century ago, the renowned ornithologist Roger Tory Peterson described a similar, though perhaps less satisfying, three days at a road-killed deer carcass that he had put out in New York State:

> I hauled [the deer] to an open slope, put up my burlap blind, camouflaged it with wild grape vines. For two days I stewed in my own sweat while the carcass, thirty feet away, ripened, and flies swarmed. The vultures, at a discreet distance, sat hunched in a tall dead hemlock like undertakers waiting to officiate at a burial. On the third day I dismantled my blind. Less than three hours later a friend chanced by; as he approached, a cloud of vultures flew up. All that remained of the deer were a few scattered bones.

Vultures are not easy to dupe, and I suspect I would have had even less luck attracting them by lying down and playing possum in the woods or meadow.

IN COMPARISON TO the body of a mouse or a bird, that of a deer can provide an advantage to scavengers such as maggots; they thrive on the soupy byproduct of bacteria, which multiply faster in warmer temperatures, and a large body that is warm at death stays warm for a long time. In short, bacteria can get a head start. I found that out the hard way, from a lawyer and a pig.

A lawyer in Boston had called me to offer what I thought was an exorbitant amount of money to serve as an expert witness on a murder case. For quite a long time, I had been researching the energetics of insects, and I often relied on measuring the cooling rate of freshly dead bumblebees in order to calculate the energy that live ones have to expend to keep warm. A bee's body temperature can drop precipitously in a minute or two, but for a pig-sized carcass, the passive cooling to air temperature, even at external temperatures near freezing, might take days, since most of the heat is deep inside the animal. In the murder case, knowing the victim's body temperature when found, which was about 98 degrees F, and extrapolating back, one could determine the time of death fairly precisely. I agreed to serve as an expert witness and decided first to collect data on the cooling rate of a human-sized pig, which would serve as an appropriate stand-in for the actual victim.

I found a pig of the right size and told the farmer I would buy it only if I could get it when freshly killed — still at the same body temperature as a human's. The limp, warm pig was delivered by truck and placed on the thin layer of snow in back of our house in Vermont. I stuck a thermometer in it and proceeded to record data. After two days the pig had cooled sufficiently for me to learn what I needed to know for the trial. But I was not going to waste all that good pork.

I butchered the pig, and we cooked some of it, but it didn't taste right — even though it was late winter and there were no flies in sight. My wife claimed that the off taste came from our chickens, which had walked with dirty feet over the pig while I was cutting it up. But I think the bad taste came as a result of bacteria growing in the long-retained body heat of this large animal, which because of the experiment had not been gutted.

To make an even longer story shorter, my ravens got to eat some hundred pounds of pork while we ate none, and although I explained my experiment to the attorney, I was not called to the witness stand (nor reimbursed for the pig or my time). That was some expensive pig. But it taught me the hard way that a large animal cools very slowly, and if you don't want it to go to the bacteria (if not also the maggots), you have to eat it or gut it immediately, because the bacteria have the inside track on their inside start.

MY PIG STORY, especially the bacteria part, leads me to creatures one step up from bacteria on the carcass food chain. As I've mentioned, within a couple of days the deer was covered with a solid layer of writhing white maggots. Given sufficient warmth, a blowfly's single "clutch" of 150 to 200 eggs hatch in as little as eight hours, achieve their full growth in three days, and complete their life cycle in a week. With their phenomenally fast reproductive and developmental rates, these insects can quickly take command. The blowflies, who had quickly discovered and tasted the carcass and found it good, had assured their colonization and dominance of it.

The large numbers of maggots and their cumulative high metabolic rates would also have *raised* the temperature inside the carcass, speeding up their growth rate even beyond that predicted by high air temperature and slow passive cooling. Probably most of the blowflies were the metallic green ones, *Lucilia sericata,* the preferred species for "maggot therapy." Maggots have traditionally been used to help heal wounds by eating necrotic tissue, and they are especially effective against Gram-positive bacteria. I suspect *L. sericata* may have chemical secretions as an offensive weapon against their main competitors for the meat, namely, bacteria. Nevertheless, I am not aware of anyone extracting chemicals from maggots the way we have extracted penicillin from *Penicillium* molds.

Maggots may be repulsive, and the stench associated with them does not endear them to us. However, we can appreciate them. They are extensively used in forensics to determine the time of death because (depending on air and body temperature) they are the first insects to colonize a corpse.

Aside from the medicinal and forensic value of the larvae is the fact that the adults can arguably be described as living jewels. A green botfly's exoskeleton is as shiny as a gem. I guarantee that if so inclined, I could sell them by the millions as ear pendants or in necklaces if they were embedded in clear plastic.

A gem is shapeless, dead, inert matter, but each green blow-fly represents exquisite aeronautical engineering; it not only flies, it has a thousand behaviors, most of which we hardly know. Like beetles, flies have no direct attachments from their power-producing muscles to the wings. As the muscles contract for the wings' downstroke, they compress the thorax and, through lever action, cause the opposite muscles — those that power the wing upstroke — to *stretch*. That stretching is what causes the muscles to contract and thus stretch the downstroke muscles. This process continues, back and forth, so that the thorax virtually vibrates like a motor and the wings beat at hundreds of times per second. In some insects, like midges, they beat more than a thousand times per second — rates impossible to achieve by direct nerve impulses alone. These flies are beautiful, inside and out.

Days after the party, more flies were still coming to the deer either despite or because of its increasing putrefaction. Flies smell (with their antennae) and can sense a carcass from about ten miles away. They walked all over the rotten meat, depositing ever more egg masses. Flies have taste receptors on the bottoms of their feet as well as on their "tongues." As the late, great insect physiologist Vincent Dethier showed, they taste sugar, salt, sour — basically the same tastes we recognize. The flies on the deer carcass were finding and lapping up proteins, which they quickly converted to eggs — proteins are the sole food of their larvae.

When no more open meat was available, the flies had won and the birds and coyotes had lost, but then there arrived another crew of carcass specialists, namely, beetles. *Nicrophorus* beetles are presumed to be able to smell a dead mouse from a half-mile away. I suspected that this insect crew might already have been working underneath the carcass, since it is well known from forensic entomological studies using pig carcasses as surrogate human ones that different species feed at different times after death. To find out if my suspicion was correct, I had to turn the deer over. It was difficult to do without gagging. But it was worth it.

The moment I lifted the deer carcass slightly, I saw dozens of long, black shiny beetles (*Necrodes surinamensis*), who have parallel ridges etched into their backs. There were also staphylinids: sleek, elongate, fast-running beetles that are commonly thought to be "short-winged" but that actually have very long wings neatly folded up like a parachute under their very short elytra. About 40,000 staphylinid species have been described worldwide, and possibly twice as many are as yet undescribed. They are mostly predacious, so it was no surprise to find them scavenging on meat. They are fast runners and good flyers, and when I lifted the carcass they scattered and buried themselves in the soil debris on the surrounding ground. There were several species. One was a shiny dark blue, and another had much yellow and brown. A third was grayish with what looked like a white stripe across the back. I noted two species of silphids, flat, broad beetles about the size of a thumbnail and mainly black. One species (*Necrophila americana*) had yellow on the thorax; the other (*Oiceoptoma noveboracense*), red.

I had expected to find many of the nicrophorid burying beetles that had so quickly visited the mouse and shrew carcasses I had set out, but after checking carefully, I found — to my very great surprise — not a one. It was hard for me to believe that they would not be attracted to deer meat. Perhaps none of the burying beetles were near that area at that time. But a road-killed chipmunk that I put out at the same time as the doe attracted two burying beetles in one afternoon. I suspect that the scent of either rot or the maggots had repelled them.

When I returned to inspect the remains of the doe two weeks later, on August 5, I found only a few tufts of hair and the stained depression where the carcass had lain. Next to that spot, though,

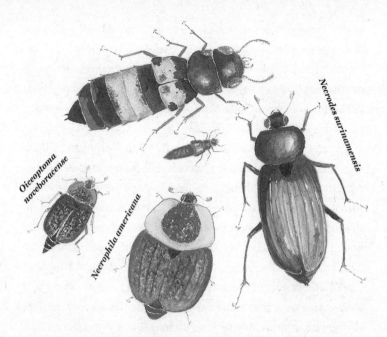

Carcass-scavenging beetles found under a deer carcass. Top: two (of more than 46,000 species identified) rove or staphylinid beetles; bottom right: Necrodes surinamensis; *bottom left,* Oiceoptoma noveboracense *and* Necrophila americana.

were two clues. The trunk of the old yellow Delicious apple tree where the vultures (and crows) had landed had the freshly grooved claw marks of a bear, and several branches of this early-maturing apple were broken. A bear had come, eaten sweet fresh apples, and dragged off the remains of the doe.

THE FOLLOWING LATE April I found a dead bull moose about two kilometers from the site where I had left the doe. A bull moose probably weighs ten times as much as a white-tailed doe. This one looked emaciated; it had apparently died from complications of moose tick disease, a common cause of death in these woods when winter temperatures are not low enough and the snows do not last long enough to suppress these parasites. If wolves had been present, there would not necessarily have been many more moose deaths, but moose weakened by the tick disease would probably have died sooner, because they would have been the predators' prime prey.

I found the moose a day or two after it had dropped in thick forest next to a small brook. Coyote tracks surrounded it, and the coyotes had chewed through its thick hide to make a hole in the throat. A raven had already fed there, leaving white droppings on the hide. Other ravens came to feed as the coyotes enlarged the hole. Later at least a dozen turkey vultures monopolized the carcass, and then the maggots "cleaned up" after them. A couple of weeks later, a pair of ravens remained after the coyotes and vultures were done; they came daily to turn the leaves surrounding the carcass, presumably picking up maggots, fly pupae, or other insects.

When all that was left was a skeleton with dried hide covering part of it, a black bear came and dragged the remains a short way downhill. After two weeks, I found little more than a pile of hair where the animal had fallen and the vertebral column and skull some distance from that. A porcupine was eating the still-fresh bone, gnawing it off in patterns that looked like a magnification of the tooth marks mice might leave on cheese. No vultures remained despite a lingering smell. Why were the vultures no longer attracted? There was no more meat, but how would they know that without checking? Do the maggots emit a scent that repels even vultures, or are they repelled by the scent from bacterial decay?

IN THE RECYCLING world of nature, there is much redundancy and always backup. The recycling process may start with a car or ticks, then employ scavenging birds, move on to flies, then beetles, and finally bacteria or, in the case of our deer and moose, a bear. Had the bear not taken the deer carcass after the flies were through, it would have been visited by swarms of dermestid beetles. These beetles, of which there are 500 to 700 species worldwide, come to a carcass when the remains are dry and well past the decomposition stage. They eat the remaining wool, feathers, gristle, fur, and skin — everything except bare bone. For this reason they are often used to clean skeletons in museums. In a forest, rodents as well as deer would gnaw on them to get their needed calcium, and the bones become covered by leaves. In my forest I find deer skulls but seldom more; the skulls always last the longest. No animals larger than moose inhabit the Maine woods. Clearly, though, the local undertakers are up to the task of disposing of even a moose in relatively short order.

But what is the undertaking process, I wondered, for something almost colossal in comparison, such as an elephant?

THE ULTIMATE RECYCLER:
REMAKING THE WORLD

The hand is the cutting edge of the mind . . .
The most powerful drive in the ascent of man
is pleasure in his own skill.

— Jacob Bronowski, *The Ascent of Man*

WE HAVE SEEN THAT MANY SPECIES ARE BOTH HUNTERS AND scavengers — the two roles share the same goals and many of the same tools. This was true for early humans: the more effective we became as scavengers, the more we hunted, and vice versa. Nowhere is that proven more dramatically than with the most challenging of prey: elephants. Humans are the only predators on earth who can consistently prey on them, open them, and also affect their existence. On land we are to the elephants what the sleeper sharks and hagfish are to the whales of the ocean depths: the ultimate recyclers. How we became that is embedded in the history of our species. Our relationship to the elephant defined our path to becoming the earth's ultimate agent for processing through the maw of our collective metabolism almost everything that grew and much that grows. Once we had learned how to access the live meat of elephants, we were equipped with the brains, the brawn, and the social organization to tackle almost anything that came within our reach. The questions I am leading up to are whether we evolved as hunters or scavengers from the start, what roles we played and still play as undertakers, and how tortoises and elephants hold the answers.

THE DEBATE ON whether we started as hunters or scavengers after we (and other apes) diverged from a common ancestor has been a heated one, and it predates a half century. My only "data point"/anecdote that bears on it, and that may bias my thinking, occurred in Kenya's Amboseli National Park around 1970, when I had the privilege of tagging along behind a troupe of human-habituated baboons with a graduate student who was studying them. The baboons were grazing, but after only a couple of hours we saw one flush a hare. It escaped the clutches of the first baboon, only to run into those of another in the large troupe. It was then quickly dispatched and monopolized by a large dominant male, although others got pieces of this prize before it was totally eaten. The "hunt" had seemed haphazard and success almost accidental, due mainly to the largeness of the group. However, another student of baboons near that time and place, Shirley C. Strum, found that baboons regularly hunt for meat.

Anthropologist Craig B. Stanford has subsequently championed the "hunting hypothesis" for early man in his book *The Hunting Apes*. He had studied the hunting behavior of chimpanzees, our closest living relatives. Chimps of some troupes regularly and systematically hunt monkeys and other prey. They eat raw meat often and with relish, consuming skin, bones, and all. They have not been observed to scavenge. Stanford points out that the males, who do most of the hunting, engage in a lot of politicking while sharing their meat. The social nature of hunting and sharing, perhaps linked to sexual privileges, may have been pivotal in our branching off from apelike ancestors to become specialized as hunters. Hunting favored social cooperation, the skills and intelligence that became human hallmarks.

From our present perspective in an industrial civilization, "hunter" is not the first term we would choose to characterize ourselves. But looking back in time only a moment, consider America in the nineteenth century. In John James Audubon's "Missouri River Journals," which he wrote between June 4 and October 24, 1843, we see a part of America that was thinly populated by a few frontiersmen and Indian tribes. On June 9 Audubon wrote: "We saw three Elks swimming across it [the Little Missouri], and the number of this fine species of deer that are about us now is almost inconceivable." On August 11: "it *is impossible to describe or even conceive* [his underlining] the vast multitudes of these animals [bison] that exist even now, and feed on these ocean-like prairies." Hardly a day went by that Audubon and his companions did not shoot several bison, deer, elk, antelope, or wolves.

It is difficult to believe that within several decades we had almost totally destroyed that world. Weaponry was important; the rifle helped exterminate the buffalo. But hunting ingenuity did not depend on rifles alone. Audubon described how many wolves were caught simply on baited fishhooks. Buffalo were chased onto ice, where they were helpless and could be stabbed. They were also caught in pens, "especially [by] the Gros Ventres, Blackfeet, and Assiniboins," who made impoundments of logs and brush with funnel-shaped passages leading into them. The bison were lured and then chased in. A young man, "very swift of foot, starts at daylight covered with a Buffalo robe and wearing a Buffalo head-dress," and he "bellows like a calf, and makes his way slowly towards the constricted part of the funnel, imitating the cry of a calf, at frequent intervals. The Buffaloes advance after the decoy," and hunters yell and advance behind them. Then all are destroyed.

The buffalo abundance would perhaps have been vastly reduced even before the Europeans came, if the various tribes — Arikaras, Sioux, Assiniboins, Gros Ventres, Blackfeet, Crow, and Mandans — had not been constantly at war with each other and thus never became as numerous as the Europeans. It might seem that no number of people could eat the vast quantities of meat from the millions of large animals killed, and in fact they didn't. The frontiersmen often took only the tongue and the warm brains and liver, which they often ate raw. As Audubon described the scene, "Now one breaks in the skull of a bull, and with bloody fingers draws out the brains and swallows them with peculiar zest; another has now reached the liver and is gobbling down enormous pieces of it; whilst, perhaps, a third, who has come to the paunch, is feeding luxuriously on some — to me — disgusting-looking offal." Audubon wrote of the searing heat, with temperatures often in the 90s and sometimes over 100 degrees. Meat would not keep fresh for hours in such heat, so a nomadic hunter had to kill daily and leave most of the meat to the wolves and the ravens.

AFTER WORLD WAR II, before coming to America, I lived with my family as refugees in a forest in northern Germany. We foraged for acorns, beechnuts, mushrooms, and berries. My father had brought rat traps, and we "hunted" small rodents with them as well as with pitfall traps. I recall how my father once trapped a mallard duck in a clever noose made with horsehair. Finding food was our primary concern, and the most memorable events of my young life in the forest of Germany were these scavenging trips.

At times when Papa and I went into the forest, it appeared to me that we were just looking around. One time — it was early spring because I remember finding a green frog on the brown leaves in a pool of water in an open beech wood — we sat down with our backs to a beech tree and were munching on a small crust of bread. It was quiet, although I suspect the chaffinches would have been singing because it was a sunny day. After a while we heard a dog barking in the distance. We listened disinterestedly at first, but then my father jumped up and ran toward the sound. I waited, and when he returned he was carrying a small roe deer on his back. He had found it dead with the dog panting beside it. He had shooed the dog off with a stick, and the prize was ours. On another occasion I found a dead boar in a spruce thicket, where it had probably died after being wounded by the British soldiers who hunted in this forest. It was winter, and I had heard ravens in that area several times on my walk to and from the village school, so I went to investigate and found the boar. It had been partially eaten, but there was still some fat on its hide. But our biggest prize came quite by accident. Marianne, my sister, who is one year younger, and I were out gathering firewood in the forest, our daily chore, when we found a dead elk lying along the bank of a small brook near some tall alder trees. The carcass was fresh, and we ran to the cabin to tell our parents, who rushed back to cover it with brush, the way cats hide their prey or ravens cache meat (as I discuss in the next chapter).

Had this been a wild and natural northern ecosystem still inhabited by wolves, hyenas, and saber-toothed tigers, instead of one transformed by humans, it is unlikely that we would have removed and eaten these prizes. The large predatory and scavenging animals would have gotten there first. Even so, it was important that we quickly remove or hide our finds from other takers. If ravens came, they could both feed on our prizes and give away their location to other human scavengers — mainly the authorities, who would have confiscated such treasures for their own use.

As a student at the University of Maine I continued to scavenge carcasses, mainly roadkill. I valued them as much as any grouse, hare, or deer I might shoot in the hunting season; every dollar I could save on groceries counted. These days I pass up eating most roadkill, though not all of it. But like my family after the war, people in many countries can't be picky. A large, fresh carcass represents many desperately needed meals, as suggested in one photograph in an old *National Geographic* of Africans harvesting an elephant carcass. A more recent photograph by David Chancellor shows a large number of people who were "trying to survive under Robert Mugabe" in Zimbabwe crowded over an elephant carcass. As Chancellor explained: "Just after dawn a villager spotted the carcass as he passed on a bicycle. It was in the middle of nowhere, but within 15 minutes hundreds of people had arrived from all directions. The women formed a ring around the elephant and the men stood inside, fighting and stabbing each other to get at the meat." The bulk of this carcass would have been quickly disposed of by mostly one species — ours. The people in this crowd had knives to open the carcass and spears or guns to back them up if the lions should crowd in to claim the carcass from them. Tools made the difference.

In the Hahnheide forest as a child, I frequently found chipped flints that had probably been used by early people. At the time I was unconsciously closely allied with the Pleistocene mindset, and now, after decades of presumed improvements of my nature by cultural conditioning, that mindset may be sublimated or redirected, but not much. I think I was then a dyed-in-the-wool predator. I was obsessed with weapons, which in my preteen years were restricted to a jackknife and a slingshot I had made from a red rubber inner tube that some kids at school had scavenged from somewhere. I was continually on the lookout for upgrades of the "perfect" twig fork and small piece of leather. Even now I feel a twinge of nostalgia when I see a forked stick perfect for a slingshot.

As a teen I became enamored of spears, bows and arrows, and then my .22-caliber bolt-action single-shot squirrel and hare rifle. Now I reserve my attention and respect for my Winchester .30-30-caliber lever-action deer rifle.

I imagine an ancient *Homo* in the Hahnheide carving a spear out of just the right kind of sapling. Indeed, when *Homo erectus* lived in northern Europe at least 400,000 years ago, "he" almost certainly carried a spear and other weaponry. Lions stalked then, and mammoths dotted the plain, but very few dead animals would be lying about free for the taking. He would have had to fight for every carcass, especially highly visible ones that could not be carried off and hidden.

THE *HOMO* LINE came out of Africa at the beginning of the Pleistocene, about 1.7 million years ago. It had evolved about a million years earlier from small, slender hominids, the so-called australopithecines, who were already making stone tools and eating large animals. How could these apelike bipedal creatures, being small and lacking the sharp claws and teeth of all the major evolved predators, have done anything other than pick up the dead and scavenge on what the much superior hunters had killed? How could they have overpowered fast antelopes, large hippos, aggressive buffalo, and elephants, all with a history of tens of millions of years of dealing with lions, leopards, and saber-toothed tigers?

One scenario is that they — and we — didn't hunt large animals. But Craig Stanford, from his perspective of studying present-day apes, and Baz Edmeades, a natural historian and self-taught archaeologist who grew up in South Africa, argue that we were mainly hunters from the start. Animals didn't usually just drop dead, and even if they did, they would not have been available to humans. A large animal is a huge meat pile, whether standing or lifeless, and in a natural ecosystem filled with its panoply of powerful predators, it would be killed and eaten as soon as it weakened. Any carcass, however it was killed, would thus potentially be controlled by powerful hunters, and they, given their prize and their already heavy investment in it, would not readily yield it to a smallish hominid already on their prey list. So if you were a hominid wanting to eat meat, you might decide that rather than facing a pride of lions you would be better off as a hunter. Even if the hominids found a freshly dead carcass, they would have to reach the meat before the big guys came. Because they lacked the large, sharp, shearing teeth of the cats, hyenas, and wolves, it was critical to have sharp cutting tools — which allowed them to kill and predisposed them to do so. In the Hahnheide, we were lucky that we did have a knife; if we had not, there is no way we could have used the carcasses we found.

But early man, to be a hunter, would also have needed other advantages to make up for his less than blinding speed as a runner. Current theory is that the advantage came from hunting in the heat of midday, thus reducing competition with the mostly nocturnal predators, and from matching and even exceeding their endurance in chasing large prey. To escape their predators (and competitors), the hominids could have climbed trees, but to catch prey both hominids and the competing carnivores could exercise their specific advantage on the ground. The cats had speed; the hominids, endurance.

Big-game animals can't hide, and they leave ample evidence of where they have been and where they are going in their easy-to-follow tracks. Their large size also predisposes them to overheating during exercise. The bipedal hominid who went naked, with a heat shield of hair on his head and shoulders and the ability to cool his body through profuse sweating, opened a niche for himself on the open plains by hunting in the daytime heat. Bipedalism and heat management allowed him to outrun his prey while also freeing his grasping hands, which had been useful in defensive climbing, to make and wield offensive weapons.

The use of tools such as rocks for throwing and poles for offense and defense generated a race for the evolution of intelligence in a self-reinforcing spiral, because undoubtedly what was crucial in the hunting game became currency in sexual selection, in the mating game. Early man's nakedness became his strength, and it unlocked the meat locker of ever swifter and larger prey. The larger the prey that he killed, the greater the social accolade. Then as now and as with the other predators, the thrill of the hunt was the proximate mechanism of success. As I will indicate later, we apparently succeeded brilliantly as hunters, even against elephants, though obviously not those existing in Africa now.

AFRICA IS SAID by some people to still have a "Pleistocene fauna," one that inspires awe. Theodore Roosevelt wrote:

> It is hard for one who has not himself seen it to realize the immense quantities of game to be found on the Kapiti Plains and the Athi Plains [near Nairobi, Kenya] and the hills that bound them. The common game of the plains, the animals of which I saw most while at Katinga and the neighborhood, were the zebra, wildebeest, hartebeest, Grant's gazelle, and "tommies" or Thomson's gazelle; the zebra and the hartebeest . . . being by far the most plentiful. Then there were impala, mountain reedbuck, duiker, steinbuck, and diminutive dikdik. As we traveled and hunted we were hardly ever out of sight of game.

This abundance, in addition to Africa's antelopes, zebras, and giraffes, as well as apes, elephants, hippos, wild dogs, leopards, cheetahs, lions, and hyenas, was long thought to be a pristine assemblage. But even though the savanna has been a feature

45

of the African continent for five to ten million years, its current occupants are only a pale semblance of what once lived there.

The easy-to-kill large animals went extinct first. Meanwhile, hominids evolved into many forms, and the australopithecines' evolutionary descendants ultimately became the large, big-brained, athletic *Homo erectus*, who mastered fire, probably talked, made stone tools for cutting and piercing, spread out of Africa, and hunted elephant-sized animals.

As the late anthropologist Paul Martin first explained, *H. erectus* were deadly hunters who spread over the globe; from one continent to another, the extinction of the magnificent megafauna followed soon after they arrived. Those extinctions include a number of species of elephants, most famously the woolly mammoths. The explorer and trophy hunter Carl Akeley, whose elephants and other African game animals are on vivid display in the American Museum of Natural History in New York, reported that the largest elephant he ever saw measured eleven feet four inches to the top of the shoulders. One of the biggest he heard of had "good-sized tusks — 80 pounds." The mammoths that *H. erectus* killed, though, were giants compared to the African elephant.

Akeley learned that killing a modern African elephant is not easy, even with an elephant gun. Nearly a hundred years ago, he was in the top of a tree in Uganda "to inspect two hundred and fifty elephants which had been chevying me about so fast that I had not a chance to see whether there were any desirable specimens [for the museum display] among them or not." Another time he was in the middle of a herd of seven hundred elephants making "a continuous roar of trumpeting, squealing, and crashing of bushes and trees." What had been jungle was trampled flat. On one occasion an old bull "took twenty-five shots of our elephant guns before he succumbed." Akeley himself almost succumbed when a bull charged, thrusting his two tusks to impale him, but they instead stuck in the ground on each side of Akeley.

It's hard to imagine how Pleistocene humans killed elephants such as mammoths with spears, but they apparently did. There are few whole adult mammoths around for comparison with African elephants, but rare specimens have turned up in Siberia (where they may have sunk into floating bogs or drowned by breaking through iced-over lakes or rivers). In 1846, during an abnormally warm summer in Siberia, people on a steamboat ascending the Indigirka River in a remote area were astounded to see "a huge black horrible mass" covered with long brown hair bobbing out of the swirling water: it was a woolly mammoth. When the carcass was dragged out of the river by horses, it was found to be thirteen feet in height and fifteen feet long, with tusks eight and a half feet in length. As the people who came upon this scene were examining the mammoth's stomach contents (fir and pine shoots along with young cones), the bank it had been hauled up on gave way and the beast was swept off in the current. Another Siberian mammoth carcass, partially eaten by bears, wolves, and foxes after it thawed out of the permafrost, had tusks that were nine and a half feet long and weighed 360 pounds — four and a half times more than the weight of good-sized African elephant tusks, according to Akeley. Did people discover a way to feast on such massive elephants? Our only direct evidence consists of some isolated spear points found implanted in carcass remains. But these mammoths no longer exist — and we have circumstantial evidence that we may have hunted some past elephant species to extinction.

The Arctic tundra mammoths (genus *Mammuthus*) lived from at least the Pliocene epoch, nearly five million years ago, almost to the last "moment" (perhaps less than 4,500 years ago), when they went extinct. Another, much smaller elephant, the mastodon (genus *Mammut*), was "only" about the size of the African bush elephant and only superficially similar to the mammoth. It had, among other differences, distinctive teeth. It was a contemporary of *Mammuthus* species living in the cold spruce, fir, and birch woodlands. Forms of *Mammut* are known from the Oligocene, thirty-four million years ago, and they, too, went extinct only several thousand years ago, again at precisely the time *Homo* arrived.

Early man's culture and hunting technologies undoubtedly developed slowly over millions of years. Significantly, as I will argue, the African elephant was probably not a significant target; it survived. The road to oblivion for the others started with much easier prey. It may have begun with a completely different animal, the tortoise, which provided a ready meal, but required a tool to get it out of the shell. Neither baboons nor chimps need tools to hunt and eat hares or monkeys, but the main human prey could not be eaten, and caught, without tools.

FIVE MILLION YEARS ago several species of giant tortoises lived in Africa. Hunting them would have been essentially identical to scavenging: whether the animal was alive or dead, the predator would turn it on its back in order to penetrate its body. There is no proof that the australopithecines did this (how could any evidence have been left?) except that by three million years ago these tortoises had disappeared. But why would the australopithecines *not* have eaten turtle? They were meat eaters, and by the late Pliocene, about two and a half million years ago, the australopithecine prehumans were leaving cut marks on bones from manufactured stone tools and were probably engaged in what Baz Edmeades calls "confrontational scavenging." Their brain size then was that of chimps' brains today. Some chimps have discovered how to extract termites from their solid mounds by inserting a long twig into a hole in the mound, pulling the stick back out, and then licking off the termites adhering to it. They pass this behavior on through their culture. The australopithecines in the Pliocene probably learned to breach the turtle's strong carapace to get at the meat within, possibly by smashing it with rocks. And smashed rocks also had edges, for cutting, or attaching to a pole for piercing.

Meat as convenient hominid food did not stop at the end of tortoise availability in Africa. It continued as the hominid line evolved to become ever-better hunters. Wilhelm Schüle, an archaeologist at Freiburg University, convincingly demonstrated that hominids were responsible for the extinction of giant tortoises by 8,000 years — an eyeblink — ago, when *Homo* reached the islands of the Mediterranean. Hominids likely exterminated most, if not all, of the giant tortoise species living where they settled. Now such tortoises remain only on a last outpost — the Galápagos Islands — and only because they were strictly protected in the nick of time. As soon as their island citadel was breached by humans, the tortoises were stacked alive upside down on ships as rations of fresh meat; in that state they lived longer than other animals would have.

For eons, hominids would have seen the tortoises as easy meat — all they had to do was pick them up. Many tortoise species lasted for millions of years simply because it took that long for the australopithecines to evolve to *Homo* and for *Homo* eventually to occupy the remotest islands of the Pacific Ocean. The major megafauna extinction in Africa did not occur until several million years after the tortoise extinctions, probably because the australopithecines were not such adept hunters of elephants and other megafauna as the *Homo erectus* who supplanted them.

A prerequisite to eating large mobile prey (whether acquired through hunting or scavenging) was having the appropriate tools. Before *Homo* came onto the African panorama, the continent hosted, as Martin and Edmeades have indicated, probably nine species of elephant-like animals, four giant hippos, giant pigs, giant wildebeest, roan and sable antelopes, giant zebra, giraffe, and giant baboons the size of gorillas. Pre-*sapiens* humans probably hunted large animals, including elephants; in Lehringen, Germany, a spear made of yew wood dating to half a million years ago was discovered with the remains of an elephant. In Boxgrove, England, a round hole, possibly a spear wound, was found in the scapula of a rhino dating to about the same time. The presumed hunters, descendants of *H. erectus* who left Africa (now commonly called *H. heidelbergensis*), had spread to Europe by a half million years ago and were making bifacial stone hand axes. They possibly used them as choppers or knives to cut up animal carcasses in what came to be called the Acheulian culture (named from tools

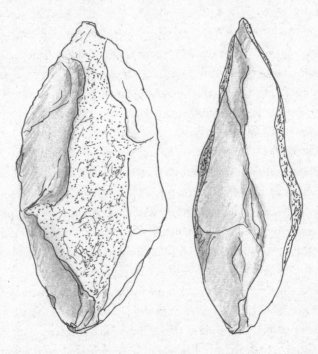

An Acheulian hand ax I found on the ground while walking in Botswana. It was probably fashioned and used by a Homo erectus *as early as 1.7 million years ago.*

first found in Acheul, France). Recent finds of these hand axes in Kenya by Christopher Lepre of Rutgers University and colleagues indicate that some of these tools date back to *Homo erectus,* 1.76 million years ago.

Perhaps the most spectacular find relating to ancient hunting tools apparently for large prey was uncovered by Hartmund Thieme in 1997 in a coal mine near the town of Schöningen, Germany. Nearly a half million years ago, Lower Paleolithic hunters living at the edge of a lake in a cool climate of spruce and birch trees left evidence of campfires and numerous stone cutting tools among the remains of many butchered animals, including horses but also elephant and deer. Their finely crafted spears were well preserved because they had become waterlogged. The spears, up to eight feet long and two inches thick, were crafted like those now used in track and field events, with their weight shifted into the forward third for aerodynamic stability. Such a find was truly miraculous, considering the near certainty that perishables such as wood and animal remains would be erased in a short time.

Another find, near what used to be an ancient lake in Kent, England, was the remains of an elephant estimated to weigh ten tons (twice as much as a modern elephant). It was apparently butchered about 400,000 years ago; its bones were surrounded by fine flint tools that had probably been used to cut it up. This elephant may have been found dead or may have been killed when it could no longer defend itself. But it might have been deliberately hunted, as had to have occurred in some cases if man was indeed the cause of the elephants' demise.

The innovation of the bow to launch an arrow gave humans a powerful tool for hunting elk and other large animals, but it could not have been effective against the seven now-extinct mammoth species or against mastodons or against the other dozen-plus elephant-like animals. Killing an elephant would have required a more powerful weapon: the spear. And if Akeley's experience killing African elephants is any indication, a man with a spear would still have been powerless facing one of the ancient elephants. However, one can make a compelling case that the African elephant is not a fair model for judging the animal that the mammoth hunters faced. The African elephants (two species are recognized now) did *not* go extinct, probably *because they evolved with man*. There would have been an arms race of coevolution: by slow stages the human hunters would improve their offensive skills, while the prey would evolve better defenses. Larger body size would have been an advantage for elephants until the hunters developed better spears. The prey then may have learned to aggregate into family groups, which might have been countered by the hunters attacking in groups, leading to the elephants aggregating into herds of many hundreds. This last step may finally have made them immune to humans equipped with only spears. Caution, aggressiveness, and cohesion to help others in the herd would also have resulted through selective pressure as a reaction to hunting by hominids over a million years, or maybe much less.

Not all of the elephants on earth had been shaped by selective pressure as a result of hominid hunting, so the people who *left* Africa entered a very different world from the more competitive one they were leaving. For many prey species, the arrival of hominids was like the smallpox virus first coming to America — or like a wildfire in dry grass.

The effect of species' isolation from humans and the consequent lack of defenses against them is demonstrated in Charles Darwin's *Diary of the Voyage of H.M.S. Beagle* (1831–1836). Here is his entry of September 17, 1835, when the *Beagle* had moved into St. Stephen's Harbor in the Galápagos, where there was an American whaling ship: "The Bay swarmed with animals: Fish, Shark & Turtles were popping their heads up in all parts. Fishing lines were soon put overboard & great numbers of fine fish 2 & even 3 feet long were caught. This sport makes all hands very merry; loud laughter & the heavy flapping of the fish are heard on every side. After dinner a party went on shore to try to catch Tortoises, but were unsuccessful . . . The Tortoise is so abundant that [a] single ship's company here caught 500–800 in a short time." A little later in the narrative, the young Darwin mentions that "little birds, within 3 or four feet, quietly hopped about the Bushes & were not frightened by stones being thrown at them. Mr. King killed one with his hat & I pushed off a branch with the end of my gun a large Hawk." Substitute mammoths for little bird or hawk, and it is almost impossible to envision such animals showing any fear of man.

Fear of predators is one of the most basic survival strategies, and animals may acquire fear behavior through genetic programming, by learning through direct experience, or by cultural learning from others. In both us and some of the other major predators, fear cuts both ways. Before we had weapons, the more powerful carnivores could have used us as prey, especially if they lured us to their kills. We needed to fear them. After we made and launched spears and poisoned arrows, they needed to fear us. We could drive them off like robbers that run at the mere sight of a cop in uniform. Until recently, when the Masai of East Africa hunted lions with spears, the lions ran away at first sight of these red-robed tribesmen. The southern African native, adventurer and writer Laurens van der Post wrote over fifty years ago how his grandfather in southern Africa told him about how the Bushmen (armed with poison arrows) could drive lions off kills with smoke and fire and then eat the rest of their kill. Anthropologist and writer Elizabeth Marshall Thomas noted how the Bushmen drove lions from their kills by an apparent accord with them. But that accord was likely cultural conditioning, derived from the Bushmen's weaponry and cleverness. Although now, armed with rifles, we might even more easily scavenge off predators' kills; pre-humans didn't have that luxury.

Fear or heightened alertness, as well as structures that hamper movement, are costly in terms of energy. It is unlikely that the giant turtles ever felt fear, because of their armor and their isolation; they just needed to retract their head into their shell. And the elephants, because of their enormous size, would not have seen humans initially as threats. A man could probably have approached an elephant to within a few meters, and a brave man could have come at it from directly under the belly, as Pygmies in Africa did until recent times.

The modern spear, the javelin used in men's track and field events, weighs 800 grams and is 2.5 meters long. It is close in appearance to the 400,000-year-old spruce spears found in Germany. The world record throw (held by Uwe Hohn of Germany) was 104.8 meters. But contemporary Olympic javelin performances likely underestimate the power of the spear as a weapon for ancient man, partly because the modern javelin for athletic events is deliberately engineered to reduce not only its range but also its power. (The world record is now, ironically, 6.32 meters *less* than it used to be, because the year after Hohn's world record was set, the governing body of the International Association of Athletic Federations instigated a change in javelin design to reduce its range.) For at least 30,000 years, hunters with spears have used an atlatl, a device that acts as an extension of the throwing arm. Sometimes an *amentum,* a leather strap attached near the spear's center of gravity, was also used to give the missile a spin, greatly increasing its accuracy and armor-penetrating power.

After we evolved to become humans and, probably armed with spears, started to leave Africa, some of us may have been nomads. Rather than settling in one place, we could continue on in hopes of more food and fewer enemies. As nomads, we had to kill for food frequently, leaving the remains to swarms of ravens in the north and to vultures in the south.

Our use of tools and our communal culture spread our cleverness, while our ability to gather as a crowd for strength when needed helped us secure animal carcasses in the face of formidable carnivores. As hunters, we would also have used animals that had died naturally. But regardless of how we obtained our meat and whether or not we caused all of the megafauna extinctions, prehistoric man ranks among the world's premier disposers of animal carcasses of the very largest land animals that ever existed.

By becoming meat eaters, via both hunting and scavenging, hominids tapped into a highly concentrated energy source. That boost in turn made even more energy available to them through evolution, first by reducing the digestive apparatus and hence reducing body weight, increasing running speed, and increasing brain size. A brain is a huge energy sink: ours eats up an estimated twenty percent of our calories, and in most animals even a one percent energy saving is a selective advantage. Any extra energy cost would be quickly selected out unless it conferred a huge selective advantage; no species will develop a large brain unless it can afford to feed it reliably. The highly nutritious protein and fat we got from animal carcasses permitted our large brain size but does not explain the reason for it, since the brains of other carnivores did not expand as much. However, relative to the carnivores, the early hominids were physically helpless, and they needed wits to take the place of what the others had, namely, armaments.

Just as chimps learned to insert a stick into a termite mound, pull it back up, and lick off the termites, early hominids probably learned that a broken rock can cut (after all, they would have experienced cuts on their bare feet). The next step was to deliberately smash a rock to give it a sharp edge and then, perhaps, fasten it to a stick to jab a half-dead animal. Killing a mammoth without being trampled may have required a half-dozen or more individuals to hurl their spears simultaneously, to drive the animal into a bog where it would get mired, or to ambush it in a ravine. Early humans were successful in these social tasks, far beyond our own good, as the extinction of at least a dozen species of elephants suggests.

As R. Dale Guthrie, emeritus zoologist at the University of Alaska, argues, Paleolithic art is a convincing indication of our ancestors' fascination, if not obsession, with large animals. They were important, and not just as food, or else these people would have depicted acorns, tubers, and beechnuts in their murals, along with horses, deer, and aurochs. Hunting required thinking, teamwork, and communication skills to create and use tools such as bows, spears, and the atlatl. Making accurate predictions necessitated intimate knowledge of animal tracks and the exercise of imagination along with constant reference to the real. Empathy for the animals would have been an almost inevitable byproduct, because the hunter needed to get "under the skin" of the animal to understand and predict its behavior. Van der Post notes that the Bushmen, who were consumate hunters, "seemed to *know* what it actually felt like to be an elephant, a lion, a steenbuck, a lizard . . ." They were well-known also for their endurance in running down prey. Along with strength, endurance, vision, and passion for the hunt, a hunter had to have sound knowledge. Hunting large animals would not have been possible without cooperation and communication among the men, but it also involved a partnership with women, for skinning, processing meat, tanning, and making clothing, tools, and shelters. The importance of the tasks at hand in the lives of the people, as in almost all animals, would have become currency for mate choice, to initiate sexual selection. A man who killed or helped kill aurochs would be preferred as a mate to one who did not do so, just as a woman who could tan a hide and fashion a warm garment from it would have been prized by a man. The ability to meet big challenges would have been a badge of worth independent of the individual's worth in other respects. As the peacock's tail demonstrates, bigger is better, despite the potential costs, and that principle is deeply ingrained. I've never met a (well-fed) Maine deer hunter who would pass up the biggest buck he could get, or one who would brag about shooting the smallest.

The cult of bagging the biggest would have been as powerful then as it is now, as it was not only a measure of success but also a basis of our livelihood. We were "successful" in eradicating the many species of moas in New Zealand and the elephant bird (which weighed up to 1,000 pounds) in Madagascar. The elephants and giant turtles had by then been long forgotten; they were not on anyone's conscience.

When the first waves of humans came to America, and the mammoths and giant sloths were eliminated, there probably was still no knowledge of limits. And when the latest invasion of humans arrived, equipped with ever new and more powerful weapons, the biggest land animal, the bison, went in a flash.

"Big" sometimes referred less to individuals than to groups. The Eskimo curlew (now extinct), called "dough-bird" because its thick layer of fat felt like a ball of dough, was shot in Nantucket in such numbers that the slaughter stopped only when the island ran out of ammunition. The passenger pigeon was doomed when the tools of electronic communication and trains took the pigeon harvesters to the immense breeding colonies. The great auk and the dodo, which also lived in great colonies on islands, became accessible by boats, and were already long gone.

THE CHALLENGE OF the elephants may have originally made us what we are, may have helped tip the balance in innovation for our eventual assault on even bigger game. Our tapping into the energy of meat some three or four million years ago instigated a "runaway" evolution, with one innovation furiously generating the next. This eventually led to a new stage of human social evolution, now cultural instead of biological, which was kick-started and maintained by the massive influx of energy from recycling the remains of 300-million-year-old plants, most prominently trees. This processing of fossil energy led to iron smelting, which opened the way to even more tools for energy extraction. Now these processes are fueling us through our farms and factories by feeding on ancient cycads, horsetails, and tree ferns. We are the ultimate scavengers of all time. Everything from the coal forests to a large part of the earth's animal biomass — domestic birds and mammals (and, increasingly, fish) — are cycled directly into *us*, instead of into a sustainable world ecosystem. So far there is little evidence of conscious effort to stop "growth." People (except possibly the Chinese) still don't viscerally realize that there are limits to human population that must be adhered to whether or not we as individuals choose to do so. Nor do we see that now the toolkit we hold is like a box of matches in the hands of a child. We'll never stop our unquenchable need for resources, but we can stop our growth, which will then provide freedom of choice in what we can use.

II

NORTH TO SOUTH

As I write today, it is mid-May in Maine. The natural world around me is dramatically different from what it was a month ago. Color has appeared as a new dimension: two weeks ago the gray-brown forested hills erupted with the red blotches of flowering red maples. A week later these were intermingled with pale yellow patches of sugar maple trees in flower, followed in a day or two by white splashes of juneberry, like white sheets hung in a rising tide of green. Their berries are still green, but they are almost always picked by a large variety of birds before they even ripen. This tree is also called serviceberry, because it blooms at the time when church services were traditionally held for people who had died over the winter; the bodies were kept aboveground until spring soft-ened the concrete-hard frozen soil.

The season of new lives' beginning and the disposition of the dead is cyclical in the north, and the flowering phenology of the trees is the best calendar of the season. But the calendar is locally specific, and burying can take place only when the undertakers are active. Here in the north, where I live, no burying beetles are about in win-ter or early spring; there is little or no bacterial decay, no flies or fly larvae, and the vultures have not yet returned from the southern areas where they spend the winter. Of the major undertakers, only some mammals and the ra-vens remain and stay active during the winter.

NORTHERN WINTER:
FOR THE BIRDS

We love the things we love for what they are.
— Robert Frost, "Hyla Brook"

OF ALL THE VARIABLES THAT AFFECT THE DISPOSITION OF
dead animals, temperature is the one with the widest implications. At low temperatures, bacteria stop dividing, insect scavengers can't fly, and vultures, with their naked necks, would freeze if
they didn't head south. When I put out the deer carcass in July, the
unusually high temperatures made it possible for the flies to arrive
in droves and win over all the other competitors, including birds
and mammals. Had the deer been deposited at the same place in
the fall, winter, or early spring, most of it would likely have been
disposed of by ravens in a "sky burial."

But the first tier of undertaker participants in northern natural
habitats are wolves, large cats, and coyotes, who may provide the
carcass by killing a weakened animal or by ripping open one that
died from hunger, age, or disease. Foxes and mustelids (wolverine,
fisher, marten, weasels) and sometimes large birds of prey such as
bald and golden eagles come next. They are followed in turn by

ravens and magpies, and finally by jays, chickadees, and perhaps woodpeckers and nuthatches, who pick up the last scraps.

ONE OF THE few places on the North American continent with a remnant of the magnificent megafauna that once existed here is Yellowstone National Park in Wyoming. Here there are now populations of deer, elk, and bison, along with their hunter-scavengers — bears, canids, mustelids, ravens, magpies, and eagles. The recently reintroduced wolf is now the top predator, although at times it is perhaps an overeager "undertaker" in that it preferentially dispatches the old and the weak, meaning that "natural" death as we think of it for ourselves is a rare event. As in all carcass disposition, there is an overlapping progression of participants. Even as the wolves are opening a fresh elk or bison carcass, crowds of ravens and magpies start arriving to take part in the feast. Individual golden and bald eagles may join in, and within a day the carcass is stripped. Yellowstone is a sample of the northern paradise that once was, where one could live in a cabin in the hills, hunt elk as needed, fish for trout, pick berries in the fall, grow a patch of garden in the summer, and later leave one's body as a token for what had been taken. Now that country, from the human standpoint, is reserved largely for viewing.

We must make do with what we have, and for me Maine is not a bad choice for a compromise of viewing and doing. It has wild woods with populations of moose, deer, and black bear, and in the last half-century, wolflike canids have returned. Ravens, the premier northern undertakers, have come back as well, because the coyotes provide the link for their survival. The coyotes open the winter-weakened and dead deer, allowing the ravens to find food in the winter, when they need to start nesting in order to have time to raise their young to self-sufficiency in a year. I love these woods. I am comfortable with them, because they are sure to outlive me, and hence so are the coyotes, the deer, and the ravens as well.

I have lived with the ravens to try to understand them, and that has meant attracting and sometimes taming them for close observation. By far the largest number of deer, moose, and other creatures wild and domesticated whose fate I and colleagues have observed in Maine's winter woods were eaten by ravens. And we still have some old raven friends who have been our guides to raven communication and other behaviors over a span of a couple of decades. Edgar Allan Poe writes in his famous poem:

> . . . I betook myself to linking
> Fancy unto fancy, thinking what this ominous bird of
>   yore —
> What this grim, ungainly, ghastly, gaunt, and ominous
>   bird of yore
> Meant in croaking "Nevermore."

Obviously Poe, describing a raven "on his chamber door," had never met one of these birds, or his had perched on his chamber door a bit too long.

IT's A TYPICAL late fall day, mid-November, in 2010. I'm at camp with my nephew hunting (but not necessarily finding, much less shooting) deer. Other friends are with us, and without all of them the days would not be nearly as cheerfully satisfying as they are. My two raven friends, Goliath and Whitefeather, are in attendance too, or at least I think they are — I can no longer positively identify them. That really doesn't matter, because if one or the other of this pair has been replaced over the last twenty years, the new bird is as worthy as the original. I have known many ravens and never met one I didn't like.

I raised Goliath from a chick in the spring of 1993. Like other baby ravens, he appeared chubby. When ready to fledge, baby ravens weigh as much as adults, but their wing and tail feathers are still quite short. Goliath was clumsy, and before he could fly he waddled in a way that might pass as a swagger, were it not for the soft and endearingly purring sounds he made when I scratched his head and he closed his eyes. He and his nest mates were later the subjects of many tests of intelligence in my aviary, solving puzzles such as stacking corn chips, carrying multiple doughnuts at once, and reaching salami suspended from a long string. There were also tests of food-caching behavior involving memory and anticipation of competitors' responses.

After Goliath grew up, he was the epitome of power and grace, as all ravens are. As his long wings sliced the air, each wing beat made a whooshing-ripping sound. Cruising at forty miles an hour on the level, he made red-tailed and broad-winged hawks look like amateur flyers; at times he would fly high into the sky and soar like an eagle on outstretched wings. Ravens are a species that can have it "both ways" — if not many ways — on the wing.

When Goliath was three years old, he was with a group of about twenty wild ravens in my Maine aviary, and he and a female from the group struck up an immediate friendship. Seeing this relationship between him and the female, Whitefeather, I gave them one section of the aviary as their own (the aviary had three sections, totaling some 400,000 square meters, built into my woods). Their section contained a shed about three meters off the ground attached under the crown of a thick spruce, simulating the roofed-over cliff cavities that ravens like to nest in. In 1996 they built a nest in that shed.

Whitefeather laid four eggs in their nest, and the pair raised two chicks as well as adopting four that I added (the offspring of Houdi, from my Vermont aviary, who had abandoned her young). I then tore out a side of the Maine aviary so that Goliath and White-feather would have access to the outside and be able to find some of the food for their young. The pair foraged on their own in the Maine forests, although Goliath was ever ready for handouts for some time, until after the young fledged. In later years, after 1996, I was living in Vermont but would go to visit the ravens on many occasions. Whenever I came near their territory, I called Goliath by name, and he would answer from the nearby forest, fly out to greet me, and accept the treats I brought him. In the summers, when my wife in those days accompanied me and we cooked our meals over an open fire next to the cabin, he usually perched in a big dead birch tree next to our grill. His mate remained reticent, hanging back out of sight in the nearby pine grove, although she sometimes called from there.

As Goliath got older and had less and less contact with us, he became more independent. I could not take him to Vermont, where I was teaching at the state university in Burlington, because he had not grown up with other people and, encountering them for the first time, would likely meet an untimely end. But I continued to visit him at regular intervals, and I always left him food. He was becoming a hunter. I found blue jay feathers and the remains of a gray squirrel under the red maple tree by the aviary, where he often perched. I presumed he was finding his own food and was no longer relying on my handouts.

One year Goliath apparently become frustrated, perhaps even angry at me for a longer than usual absence. On occasion he had been inside the cabin with me. Perhaps he thought I was inside and was refusing to come out to feed him, though of course I have no idea what he was thinking. But I do know what he did: when I came back to the cabin, I found a lot of the chinking between the logs pulled out, although Goliath had never taken a fiber of it before. He had also visited the outhouse, where he had taken a roll of toilet paper, unrolled it, and left tatters all around in the trees and on the ground. After that he never again came to me. For all practical purposes, he might as well have left, and I thought he had.

I was absent from Maine most of the fall of 1997, when Goliath was four years old, and when I did go there I saw neither Goliath nor Whitefeather. However, I returned right after New Year's in 1998 (the winter of the great ice storm in Maine) to meet with my winter ecology class a few days later. I saw no ravens for almost two weeks, but on January 10, as we were about to leave, the pair suddenly arrived as if out of nowhere. I was greatly surprised to see them. They were both making a big racket at the aviary and inside the shed where they had nested.

They were getting ready to nest again. Like other male ravens, Goliath wanted to return to a nest site that had proven successful. He entered the aviary repeatedly to inspect their old nest. She, however, refused to enter. I had not seen him for eight months, and he seemed uninterested in me, but he was willing to perch directly above me on his favorite perch in the red maple, where he had occasionally left scraps of uneaten squirrel. He kept insisting on returning to the nest, and she kept resisting. Finally, in April, very late for local nesting, they "compromised" and built their nest high in a nearby large pine tree; on May 8 she was incubating the eggs — at a time when most other local ravens have ready-to-fledge young. After I left two white chicken eggs on a stump under their nest tree, I heard them make a big vocal commotion, and then they seemed to vanish, abandoning their nest; I didn't see them for two days. Eggs are a favorite treat of ravens, but they don't normally find them lying out on a stump. I don't know what they thought about the eggs, but it apparently led them to reject their nest.

I climbed up to the nest and found their four eggs covered with the nest lining of moss and deer hair. The ravens never returned.

A mated pair of common ravens, Corvus corax, *in portrait and while preening each other; the birds form lifelong pairs.*

This was very unusual. I have examined dozens of raven nests containing eggs, and in no other case was a nest abandoned, even after I had inspected it repeatedly and even after I added chicken eggs of natural color or painted bright red or green as well as flashlight batteries, potatoes, or rocks; all my additions were accepted and incubated. I have no idea why my estranged (from me) pair left their nest immediately after my generous offering, whereas a wild pair would not have done so. At one time Goliath had expected me to be a continual food provider, and after I abruptly stopped provisioning him, he savaged my house and abandoned me. A raven never forgets; it seemed I was now worse than a thief. When I suddenly offered them the one delicacy they desired most, perhaps they thought it a trap — I could only be up to no good.

I was afraid the birds would now leave their territory. However, although they did not nest that year, they apparently settled their "argument" (or there was a divorce), because most years since then they (or another pair) have nested in the same pines, always starting early in the season and finishing the job.

Disagreements between a mated raven pair may not be unique. Near my home in Vermont the local pair built their nest in 2009 on a small cliff, but a predator got the young shortly after they feathered out. The next spring the pair started a nest there again, but after it was half constructed they abandoned it and built another to completion in a nearby pine tree. There they successfully raised a brood. In the spring of 2011, they built a partial nest on the same branches of the pine tree, then abandoned it and rebuilt a second nest on the old cliff site, from which they later fledged their brood.

ON THIS NOVEMBER day in 2010, I hear the raven pair call from their sleeping (and nesting) place in the pines near my Maine cabin at dawn, as they do almost every day year-round. I cannot tell which one is saying what, but they often go through a repertoire of several different kinds of calls, and after several minutes of "conversation," they take wing, flying either together or separately. In the evening they return. I never know where they have been, but when I am out in the nearby forest during the day I almost always hear raven calls from somewhere; sometimes a single bird flies over, sometimes a pair. I never recognize them, but one could be Goliath or Whitefeather. (The colored plastic ring that Goliath once wore has certainly by now worn through and fallen off. Whitefeather's wing tag was attached by a metal rivet, and the cold metal caused a nearby feather to grow back white; after the tag was lost, the feather, replaced in the next molt, may have returned to its original color.)

Today Charlie and I hope to be able to leave them something — namely, the entrails of a deer, if we can get one. Two years ago, when we were successful on the hunt, we left a gut pile about a half-kilometer from the cabin, and ravens were on it within an hour. Today, as I'm perched with my rifle high in a spruce tree at 7 A.M., I hear the *swoosh, swoosh, swoosh* of a raven's powerful wing strokes. No other bird that I know makes such a sound, one evincing such physical power. The bird flew directly over me, showing no sign of recognition. It flew on and then stopped nearby; for the next half-hour I heard a continuous monologue of quarks and gurgles in all sorts of cadences, pitches, and intonations. This raven "song" consisted of bell-like sounds and a happy-sounding, high-pitched gurgling trill that I had heard before when I'd left them an attractive carcass, such as that of a deer. I didn't know if the bird was happy, but I was certain it was not depressed — it was perhaps anticipating great meals. I wondered if the raven had spied a deer or moose and, given the hunters' (our) presence, associated it with food. If so, the mere expression of its joy could have been a self-fulfilling prophecy, provided that a nearby hunter knows a raven is happy and makes these sounds when it sees a deer or other potential food.

The ravens' music continued for twenty minutes, until Charlie joined me as we had planned.

"Did you hear the raven?" I asked.

"Sure did. I went right by there, and I also saw a very fresh deer track near there." A half-hour later we got our deer.

The raven's calling is not likely a conscious act "to" accomplish an objective. As much as I think the songs of the thrush, warbler, and finch are expressions of joy and vigor, I "know" also that they function "to" attract a mate, "to" proclaim a territory, and "to" repel rivals. Unfortunately, such knowledge, which allows me to assign specific functions to the calls, lends these animals a somewhat mechanical aspect in my own mind. Yet in the raven, it is not necessarily objective to make such interpretations.

What I had heard was like a jazz improvisation, but not by just one instrument. It sounded like a medley of many voices. The raven seemed to be having fun. I can't think of a cheerier bird song, except possibly that of the winter wren. But if the wren is singing for fun, it's fun for only a short time in the spring, just before nesting. The raven sings, though rarely, at any time of the year. It is play.

Play is a raven characteristic. And play is an expression of "pure" joy. It does not require a reward other than the doing of it. You see it also in the raven's flight. At all times of the year, you can see a single raven flying along steadily in a straight line, as though it has a destination, and I presume it does. The bird, apparently absorbed in some monologue of gurgling or other calls, may suddenly tuck one wing to the side and tumble down like a black bomb in a corkscrew dive, catching itself a couple of seconds later with an outstretched wing, only to vault back up and continue its straight flight just as before. It looks like plain exuberance. And this behavior usually has no raven audience.

I do not know what a raven's "spirit" is. But if I had to choose this bird's defining characteristic, it would be the opposite in many ways of what has been depicted in popular culture. Far from being "ghastly" and "grim" à la Poe, ravens are the cheeriest birds on earth, especially near a potential feast, and what's more, they perform sky burial most joyfully. If I could choose, I'd be reincarnated into the raven.

Ravens are, or at least were, arguably the most important carcass consumers in the Northern Hemisphere. They are the premier carcass specialists of the crow family (magpies in winter may be the second), although they cannot feed on a carcass until the mammals with sharp cutting teeth have opened it. Normally the mammals come first to winter-killed animals, and the ravens follow. In fieldwork in the Maine woods, my postdoc John Marzluff and I found that ravens disposed not only of moose and deer but also of about two hundred stillborn calves, numerous goats, sheep, cows, and horses, and all sorts of roadkill, from raccoons to porcupines.

We never saw large numbers of ravens except at carcasses or near communal roosts whose members slept at night and then fed at nearby carcasses the next day. When we did see numerous ravens at one time there were usually less than fifty. However, after we wing-tagged more than four hundred so that we could identify

Ravens feeding at an elk carcass along with the carcass openers, their providers.

individuals, we saw that there was an almost continual turnover of birds at any one carcass.

IN THE WINTER, when the ground is stone-hard from frost, the insect undertakers are inactive and the winter crew of warm-blooded animals, mainly coyotes, cats, foxes, and ravens, come in as their substitute. The cold meat stays fresh, so it can be a prized resource potentially for months, and the longer it is available the more birds it seems to attract. I once obtained two full-size skinned Holstein cows, each weighing about a ton. I wanted to find out if it was physically possible to satisfy the ravens' appetites and perhaps overwhelm their capacity to devour a carcass. From the numbered and colored wing tags of individuals that could be identified, and then from the number of these marked ravens that came to the feast, I was able to calculate the total number that fed there: close to five hundred individuals. In two weeks the ravens had removed almost all the meat from the two cows. That does not mean, though, that they had eaten it all — far from it.

To counter the rapid depletion of the carcass, the birds try to haul away as much as they can to cache for future use. They fly off with a load of meat, land on the snow or ground, lay the food down, then use their bill to excavate a small hole They put the meat into the hole and cover it with snow or nearby debris, then quickly fly back to get another load to hide somewhere else.

Since this is winter work, and the carcass is often frozen rock-solid, it takes much time and effort to chip off meat. In subzero temperatures, getting the meat off the carcass can take as much energy as the meat itself supplies, so the birds try to steal from others by watching where they make their caches. As a counter-strategy to *that* strategy, the cache makers remove themselves far from the sight of potential thieves. In Maine, at winter carcasses where a crowd of birds has gathered, streams of them fly off in all directions making caches. Unless the carcass is defended and used by only a mated pair, the birds usually fly a kilometer or more to make each cache. Each bird makes lots of them, one after another. They need to, because much of the meat will be found by those from whom it can't be hidden, namely, the mammals who hunt by scent. I suspect that canids such as coyotes, who are afraid to approach carcasses tainted with human scent, may rely on the ravens' caches, in a kind of quid pro quo for having provided initial access to it.

In Vermont I have built a platform about ten feet off the ground as my raven feeder. All roadkills and kitchen scraps go onto it, and a local raven pair now "own" it. I believe our dog, a yellow retriever named Hugo, thinks the ravens are his providers, because he often rushes down to my "raven restaurant" after he has watched through the window as they arrive. He occasionally finds a scrap of food they have dropped. He also robs their nearby caches.

I do not know how much of the raven-cached meat is recovered, either by the ravens who cached it or by other animals, but I suspect it could be a lot. Meat has scent, and Hugo seems to find the ravens' caches quite easily, even those he did not observe being made.

When ravens feed at a carcass in the winter, there will soon be meat scattered for kilometers around. Much of it will be taken by coyotes, weasels, deer mice, red-backed voles, flying and red squirrels, fishers, and both short-tailed shrews (*Blarina*) and common shrews (*Sorex*). Thus any carcass that ravens feed on in the winter will be recycled, not only into the birds but also into a large portion of the mammal fauna that needs meat to survive the frozen season.

NOTE: AS THIS book was going to press in November 2011, a lone raven with a white feather in its right wing was sighted on the ground in a clearing near my cabin. This put a new wrinkle in my story about Whitefeather, especially because during that same period I had seen some very unusual raven behavior. I awoke on November 9 from solid sleep about 5:30 A.M., hearing agitation calls from "my" pair of ravens at their usual roost site in the pines next to my cabin. As I stepped outside into the early morning darkness, the pair lifted off and began to circle the clearing. A third bird continued to call from the pines, then joined the pair but soon returned to the trees. The pair in the air kept up a continuous conversation of nasal hanks and honks, squeals, *kecks*, hooting, and knocking calls (the female courting calls). The calls got fainter and fainter to my ears as the birds circled perhaps more than a mile above me. The third bird remained perched, occasionally calling loudly.

The pair were still cavorting after it was fully light; I saw them as barely visible black specks dancing above the clouds. I stood spellbound, continuing to take notes on what I later wrote was "an incredible raven experience — I was almost breathless with wonder — I've never experienced anything like this."

They performed their sky dance, during which I never saw them more than three feet apart, for at least an hour before bursting through the clouds like black thunderbolts. They pulled in their wings and shot almost straight down, then caught the air again, gracefully spiraling up and again falling like rocks. It was a ballet of sound and motion.

At noon and then at 4:20 P.M., when it was almost dark, I again observed their dance. At dark they returned to their sleeping place in the pines, but this time I heard the aggressive, staccato *keck* calls, accompanied by a vigorous, straight-out chase, the birds skimming directly above the treetops. A third bird followed, seemingly passively.

The next morning at 5:30, with the stars still shining brightly, the ravens again woke me, and I jumped out of bed. They repeated their aerial display. The third bird stayed in the trees in the vicinity, occasionally making the long, undulating territorial calls that I have come to believe are unique to the ravens who own this hill. The pair stayed aloft for forty-five minutes before returning to earth. Again they interacted with a third bird, although I could not determine what was going on because they were flying through the forest. I was never close enough — or above them — to determine whether one was Whitefeather, though I suspected so.

Without being able to identify the individuals or to follow them through the vastness of their domain, I can't possibly say what was going on. Except for one thing: I may have made an error of interpretation due to an assumption.

In January 1998 I had assumed that it was Whitefeather who returned with Goliath to their old nesting site, the aviary. I now think it equally likely that there had been a divorce, and the new bride reluctant to enter the aviary was in fact not Whitefeather, who might have returned to her *old* territory on release. Perhaps now, thirteen years later, she had, for any of several reasons, come back. Perhaps Goliath was now trying to reaffirm his old bond with her, thus pitting his new mate against the old one, as each contested for the same mate and the same territory. Given what we know, there are several possibilities; many tales could be spun.

THE VULTURE CROWD

My first memorable encounter with vultures occurred when I was twenty-one years old and in Tanganyika (now
Tanzania), East Africa. I was exploring "the bush" at the edge of
Dar es Salaam. Here's an edited version of what I wrote in my
journal:

October 24, 1961. The Africans here keep large herds of the
multi-colored humpback cows. At night they are held in corrals
of bushes and thorns but during the day the herdsmen drive
them at a slow pace through the countryside. These cows all
look rather skinny, and I think they are butchered only when
sleeping sickness or some other disease overtakes them and
they drop in their tracks. This morning I came across a freshly
skinned carcass from which most of the meat had been hacked
off. It was on the bottom of a narrow ravine, and could not have
been visible from afar. It had not yet been touched by animals
as it was just getting light (early in morning) when I came upon
it. I returned to the spot an hour or so later and sat and watched
and listened for a half-hour. At first only a few vultures perched
in the surrounding trees. Some of them came closer to the car-

cass, and the trees they left were promptly occupied by others, who literally dropped themselves into the branches so that they shook and rattled. Then one or two dropped down to the carcass. Then, as if by signal, vultures were suddenly swooping in from all sides, and now they were no longer bothering to settle in the trees but instead came straight down into the ravine, plummeting on motionless wings past where I sat on the edge. The wind rushing through the wing feathers produced a flopping, droning sound, like a flag in a hurricane. As more vultures swooped down, more came in. In the distance I saw black specks that were formations of still more vultures that could not possibly see what was going on down below me in the gorge. They arrived in surprisingly short time, broke ranks, circled once or twice, and rushed down to the ground with outstretched legs. In half an hour there must have been 150 at the carcass. There was a large mass of them on top of each other and fighting from the sides to get in. It was all rather quiet except for the thrashing. Occasionally there was a squeak or a screech as two faced each other with outstretched naked necks and wings outside the squirming mass. After a while, some flew off with laborious wing-beats and settled in nearby trees. I saw others spiraling on motionless wings high in the sky. Swarms of them, with ever more arriving as if from nowhere — first as barely visible specks in the sky — until they overpowered my senses.

In some places such scenes are still current. The wildlife biologist Richard Estes, leading a safari in Namibia in January 2011, told me of seeing about a hundred vultures around a recently dead giraffe carcass. When he went back to it an hour later, the carcass was still intact, but by then there were nearly three hundred vultures, apparently waiting for lions or hyenas to provide access to the meat. At about that same time, a visitor to the city of Dakar,

Senegal (on the Atlantic coast of West Africa), told me of seeing "swarms" of vultures flying all over the city and "hopping along roadsides." Pied crows, which in Africa take the place of ravens as scavengers, were in equal abundance. Many horses and goats roamed apparently free in town, along with dogs and cats. It is not hard to imagine the scene wherever a goat, horse, or cow dies or is butchered in an area where vultures live; there would be no scrap of offal left after several hours.

UNDERTAKING SURELY IS an ancient heritage. Although undertakers were, and still are, often not differentiated from executioners, they have always been the essential link for the continuity of life; without them, life would have come to a grinding halt. Over millions of years, through evolution, the body size of herbivores and predators and their handmaidens, the scavengers, increased. As the herbivores grew larger, so could those that took advantage of their fallen bodies. For every individual that walks, one dies, and each one becomes a resource of highly concentrated food. The larger the carcass, the more food there is for those who feed on it. In turn, a large amount of food in one place for a short time favors larger scavengers, because it tides them over from one meal to the next.

Consider the apatosaurs of the Cretaceous, 145 to 65 million years ago, the largest land animals that ever lived. Such huge meat piles, weighing as much as thirty-eight tons each (the equivalent of eight or ten African elephants), left regularly on the landscape, would have been used. The larger the meat pile, the more it was worth defending, and strong defense would also have favored large size. The tyrannosaurs, with their bulk of up to nine tons, were not suited for running pursuit and agile maneuvering. Their long, sharp teeth were adapted for ripping flesh, though, so they may well have been in the first tier of undertakers. Along with feeding on the already dead when opportunity afforded, they would have dispatched the old, the weak, and the injured. Yet a tyrannosaur would not have picked clean an animal as large as an apatosaur; there would have been plenty of scraps left over for many more tiers of undertakers. And, since carcasses of very large animals were by definition not numerous (a plant community can support only a certain amount of biomass), the long distances between potential meals would have favored the evolution not only of huge undertakers but also of some large flying ones.

The gigantism of the Cretaceous herbivores probably explains why the largest known flying creatures, the pterosaurs, lived at that time as well. The largest of all, *Quetzalcoatlus* and *Hatzegopteryx*, had wingspans of at least ten meters and maybe up to twelve meters (the largest vulture today, the Andean condor, has a wingspan of three meters). Their size, which would have greatly compromised their maneuverability and ability to hunt, would have been an advantage both in fighting for carcasses and in getting to and surviving on infrequent but large meals. We can infer that these pterosaurs played a major role as the cleanup crew that swooped in to feed on the giant carcasses after the meat-eating dinosaurs had penetrated the hides and perhaps taken most of the easily accessible meat of their large herbivore prey. The pterosaurs would have been super-vultures.

These giant meat-eating scavengers, specialized to travel far and to live off, and hence to require, huge carcasses, went extinct at the end of the Cretaceous when an asteroid struck the earth, causing climate changes that wiped out their food base. The small and relatively inactive reptiles, such as snakes, turtles, and crocodilians, who could live without food for months to perhaps a year or more, survived. But for some reason that is shrouded in mystery, one surviving line of small dinosaurs evolved to become what we now call birds. Some of the largest of these are the present-day vultures, who evolved as specialized undertakers of very large carcasses under the selective pressures postulated above for their ancient predecessors and ecological counterparts.

Among the survivors of the Cretaceous asteroid impact were the first mammals, which at that time were small and inconspicuous. Over the next millions of years, some of them — as before, especially the herbivores — evolved to gigantism and occupied the niche that the large dinosaurs had held. By the late Miocene (the Age of Mammals), some six to eight million years ago, there existed a large assemblage of huge mammals, some of whom resembled our modern fauna. The mammoths, mastodons, giant beavers, glyptodons, giant ground sloths, and other giants existed almost to the end of the last ice age. Humans were for a while contemporary with this fauna. In a sense, the reptilian actors on the stage had been largely replaced by mammalian counterparts, and large birds had replaced the giant flying pterosaurs. At that time lived the largest flying bird ever discovered, *Argentavis magnificens,* commonly known as the giant teratorn. It had a wingspan of 6–8 meters (19–26 feet) and is estimated to have weighed 60–120 kilograms (140–240 pounds). The Andean condor is a midget in comparison. It is safe to guess that a large flying bird such as the giant teratorn had the habits of a condor or vulture, especially since its large, slender beak with a hook at the end was ideally shaped to tear flesh from the carcasses provided or opened by the dire wolves and great saber-toothed cats.

Some of the teratorns persisted into the Pleistocene, and they, along with ravens, would have been well known by early man. One recent species, *Teratornis merriami,* was found in 10,000-year-old deposits in the La Brea Tar Pits in Los Angeles, along with dire wolves, saber-toothed cats, mastodons, and giant sloths — all now extinct. This teratorn had a wingspan of 4 meters (11.5–12.5

feet) and weighed around 15 kilograms (33 pounds); a California condor weighs about 20 pounds. It would have been a contemporary of the Paleo-Indians, such as the Clovis hunters, who preyed on the North American megafauna. With their extinction, that teratorn also disappeared, as did another of the largest avian undertakers, *Aiolornis* (originally *Teratornis*) *incredibilis*, which was first described from only a single individual, discovered in Smith Creek Cave, Nevada. The bird probably had a wingspan of 16 to 20 feet. Present-day vultures regularly fly a hundred miles to feed. This teratorn surely had an even greater flight range.

There are other, generally large, carcass-scavenging birds. The vulture habit evolved in the South American caracara (a falcon) and in the African marabou stork, *Leptoptilos crumeniferus*, which has converged on vultures in having a naked head and neck, which greatly helps solve or eliminates the problem of feather hygiene. To some extent the vulture habit exists also in eagles; the bald eagle feeds on dead fish and other offal, as does the bearded vulture, *Gypaetus barbatus*, a Eurasian eagle that lives almost exclusively on carrion.

Vultures evolved at least twice and arguably several times. The present-day "true" vultures are divided into seven New World and fifteen Old World species. The vultures of Asia, Europe, and Africa belong to the family Accipitridae and are thought to be relatives of hawks and eagles. Some, like the griffon vulture, *Gyps fulvus*, have naked heads. Although the bearded vulture has a fully feathered head and eats mostly carrion, it prefers fresh meat and specializes on bone marrow; it shatters large bones by dropping them from great heights onto rocks. Many of the vultures of North and South America, family Cathartidae, are similar in appearance to those Accipitridae that specialize in rotting meat and have naked heads. Resemblances between the two groups result from convergent evolution, with similar adaptations favoring the same feeding habit. The Cathartidae may be derived from storklike birds; their origin is still debated.

Vultures, regardless of taxonomic affiliation, are specialized carrion scavengers of large animals. Many prefer fresh meat, and some, like the black vulture of the Americas, will, like ravens, hunt live prey. Few other scavengers (such as wild pigs, dogs, and eagles) can compete with them for decomposed meat because they are able to metabolize (detoxify) natural bacterial biotoxins. Their habit of feeding on rotting carrion in warmer regions seems to work to these vultures' advantage, perhaps in part because it makes them taste bad; few animals eat them. Indeed, they use their partially digested food in projectile vomit as a defense, and if that tactic alone doesn't work, some "play possum," which is presumably most effective if their feathers are soiled and they smell appropriately necrotic after a rotten meal.

Vultures tend to be social, often roosting, and sometimes nesting, in colonies. This predisposition allows them to bond with people, and they are said to make good pets. A friend of mine, who was bonded to an Andean condor (and vice versa), traveled with it in his van and let it fly free on occasion, to show off the bird's awesome size and beauty. The van was its cave, to which it returned and where it roosted contentedly as long as it was well fed on fresh meat.

Vultures live primarily in warmer climates. When several species coexist, as they do in Africa, "guilds" consisting of several species use carcasses cooperatively, with each species specializing; mutual dependence then commonly results. In the Americas, for example, a turkey vulture is usually one of the first vultures to arrive at a hidden carcass, which it can detect by scent. Other vultures, lacking a strong sense of smell, find the carcass by following the turkey vulture. The turkey vulture, however, being relatively small, cannot tear into a large carcass. The larger vultures, such as the Andean condor, make the meat available to them, but at the cost to the turkey vulture that the larger birds feed first.

Vultures, not having to chase and capture prey, tend to be slow and to have low metabolism. They find food by soaring, which has nearly the same metabolic cost as perching — it's the equivalent of perching in the sky. But soaring requires warm updrafts. At night the vultures' body temperature drops, saving more energy. Because of their large size and because they have a crop in which to store food, allowing them to gorge when and where a quantity of food is available, they are also adapted to survive fasts of several weeks or more.

The vultures' conservative lifestyle is reflected in their natural history. They take a long time to reach sexual maturity — six years in the larger species. They live long; Andean condors have a lifespan of at least fifty years, and a griffon vulture has lived forty years in captivity. Having adapted to a low natural mortality rate, they have a correspondingly low reproductive rate. The larger species breed only once every two years and have only one chick per clutch, while the smaller species may have two.

NOWHERE IN A NATURAL ecosystem is the task of carcass creation and disposal more out in the open to human eyes than in the Serengeti region of East Africa, an intact ecosystem with a virtually Ice Age fauna. Six species of vultures live there. Some 12 million kilograms (the weight of about 200,000 men) of soft tissue (meat) per year is available for vultures in the Serengeti, and the birds find almost all carcasses, even those hidden in thick brush.

The specific scenarios of undertakers of large carcasses in Africa vary, but the pattern is like the following partial sequence I observed in 1995 at the carcass of an adult reticulated giraffe, *Giraffa camelopardalis*, the largest ruminant and also the tallest land animal. Bulls may be 19 feet tall and weigh more than 4,000 pounds. There is a lot of meat on the platter for scavengers when one of these animals drops.

The giraffe I observed had lived in the acacia bush in South Africa's Kruger National Park. It had probably been down only one day when I found it lying within a hundred meters of a sandy road in semidesert acacia gallery forest. It was late morning, and some of the show was already over, so what I write here is largely extrapolation and speculation.

I suspect that the giraffe was old or sick, because a healthy one is not normally taken down by lions, several of whom were nearby. These cats, weighing 250 to 300 pounds, can ingest 35 pounds each in a night of feasting. In the heat of the day, they were lying in the shade of an acacia tree.

The lions had probably killed the giraffe in the night, and their commotion attracted hyenas and jackals. After satisfying their appetites, the lions yielded to the insistent harassment of the hyenas, who, when they in turn were sated, yielded to jackals.

Soon after the morning sun had warmed the plain and the warm air had risen, vultures flew from their perches in some communal roosting place. They spiraled ever higher, their sharp eyes scouring the plain. The first vultures to see the carcass and the lions, hyenas, and jackals scattered about stopped soaring and started their gliding descent. Other birds in the distance, also soaring, watching not just ground but also sky, saw the first vultures begin their descent and did the same. And so on, one vulture informed by the next more distant one, until a horde of hundreds was barreling in from all directions, perhaps from over a hundred miles away. Some of the vultures were already done feeding when I arrived. They were perched in the nearby trees, and some were flapping off again.

In a day or so not much would be left of the carcass. Any meat remaining would be flyblown, with writhing masses of maggots.

After the large meat eaters leave, the remains are reduced to dry bones and hide and fur, and then beetles fly in, and they and their larvae finish up the scraps.

Meanwhile, the lions, hyenas, and jackals process the giraffe's remains into scat, and (as I will explore later) dung beetles process even that last remnant of the giraffe, including whatever it had eaten during its lifetime. The beetles fashion animal dung into round balls, which they roll for long distances and then bury, to serve as food for their offspring. After the next rains, when the earth softens and fresh green grass and flowers spring up anew, the young beetles emerge at night and fly off, skimming over the veldt, guided by the scent of antelope, elephant, rhino, hyena, or lion excrement, in search of further feasts. As they fly, many are caught and eaten by bats at night and by gorgeous birds (including fishers, drongos, starlings, and rollers) during the day. One giraffe died, but a dozen lions, hyenas, and jackals and perhaps hundreds of vultures were fed. Thousands of dung beetles had a feast, and the plains would grow more grass.

WE KNOW LITTLE about animal undertaking in the natural eco-system of North America in the Pleistocene. We have glimpses, though, and there is room for speculation. In 1953 the renowned ornithologists Roger Tory Peterson (an American) and James Fisher (British) got together to take a 30,000-mile journey to "Wild America" and write a book about it. In the prologue, Peterson wrote that they were both "completely devoted to the study of the natural world — the real world." The highlight of their trek was seeing a single American condor in the distance in California. The condor, Peterson wrote,

> was like a bomber, its flat-winged posture quite unlike the glider-dihedral of the turkey vulture. It was huge, black, pale-

headed, and as it came over, the big white bands forward of its under-wings showed it to be an adult. For five minutes we watched its monstrous ten-foot span, its primaries spread like fingers. It made a couple of flaps, as if it had all the time in the world, caught a new thermal, and soared away to the southeast until it became a tiny speck and disappeared.

The sad part is that this disappearance, then as now, seemed literal. Peterson and Fisher were skeptical that the California condors could survive as a species, and they wondered whether putting out carcasses might help the remnant population, then estimated to be about sixty individuals in the whole world, survive. A "last-ditch effort" had already been made to save the birds by artificial propagation, following the example of Andean condors, which had been bred in captivity at the San Diego Zoo. Normally in the wild, under ideal conditions, condors raise only one chick every two years. But in the zoo it was possible to get four offspring from a captive pair over that same time by taking their first egg to hatch in an incubator. The female immediately lays a replacement egg, which she is allowed to keep and raise, but only until the chick is half-grown. It is then removed and hand-raised, at which time the female initiates a second breeding, yielding two more chicks by the same process of egg removal and hand-rearing.

Peterson had suggested such a plan for the California condor, but it had unfortunately not worked out. Permits for the project had been obtained from the California Fish and Game Commission, but the attempt to capture a pair of birds for captive rearing had failed. When the permit expired, the project was abandoned. Fortunately, the "last-ditch effort" was again tried, though not without controversy, thirty-four years after Peterson suggested it.

Not until fourteen years after Peterson and Fisher's 1953 trip and their dire prognosis was the California condor placed on the

federal endangered species list. Even then their numbers continued to spiral downward. Eventually only twenty-two individuals
remained of the entire population, so in 1987 the controversial
decision, by then truly a last-ditch effort, was made to capture *all*
of these wild birds and bring them into captivity. This time the
capture attempts were successful, and the last remaining wild bird
was caught on April 19, 1987.

The goal of the U.S. Fish & Wildlife California Condor Recovery Program was "at a minimum to establish two wild-breeding
populations of at least fifteen pairs each from the captive-hatched
released condors." Male and female condors look the same; pairs
mate for life, and the mates take turns incubating their one egg
for fifty-six days. The young fledge after six months and may still
be dependent for another half year. Some of the captive-bred individuals, which were released into the wild beginning in 1992, may
still be alive, since these birds live up to sixty years, commensurate
with their very slow reproductive rate.

As might be expected in such a worthwhile, major, but risky undertaking, the captive-rearing program ran into difficulties. Five
of the released birds soon died from lethal contacts with electric
power lines. A program was then initiated to train the remaining
captives to avoid power lines by administering small shocks if they
landed on one, before releasing them into "the wild."

Currently there are three condor release sites, one in California,
another in Arizona, and a third in Baja California. As of 2011, the
world population of California condors totaled 369. Of these, 191
are in the wild (97 in California, 74 in Arizona, and 20 in Baja),
and the rest are still captive. The world's largest vultures, the Andean condor, *Vultur gryphus,* and the Eurasian bearded vulture,
are designated "near threatened" by the International Union for
Conservation of Nature.

Prehistorically, the California condor lived in the West from

Portraits of the North American vultures: black vulture (left); turkey vulture (top right); and California condor (bottom right). Note the flight silhouettes of the black and turkey vultures.

Canada to Mexico and in the East from Florida to New York. Its bones and eggshells have been found in nesting caves in the Grand Canyon. But 10,000 years ago it experienced a dramatic reduction when man came on the scene and the mammoths, giant sloths, and saber-toothed cats went extinct. At that point the condor was probably restricted to the Pacific coast, where it apparently scavenged on the washed-up carcasses of large marine mammals, as the explorers Lewis and Clark witnessed. After the influx of Europeans, the birds suffered from habitat destruction, DDT, and lead poisoning. Sadly, "the wild" is no longer a suitable place for some vultures.

• • •

WE ARE IN a human-generated wave of animal extinctions, and animal undertakers are especially hard hit by it. The greatest cause of population crashes, and possible extinctions, of some of our large undertaker animals, such as the great cats, hyenas, wolves, and vultures, is our decimation of the vast herds of ungulates, which provided their food base. In addition, we have traditionally granted the large undertakers little respect; killing the scavengers that feed on dead animals has been encouraged in part because they have often been blamed as the killers. As I have emphasized, the line between predator and scavenger is a thin one in some species. Both predators and scavengers are hated, especially by herdsmen whose livelihood depends on domesticated animals selected for their tameness and helplessness, who, like sick animals, are apt to become prey. Herdsmen tend to see an animal death not as a recycling into other life but, if perpetrated by an animal other than themselves, as a criminal taking of life and their livelihood. They see predators and scavengers as direct competitors and hence deserving of retaliation. Many predators are easily killed by being lured to a tethered live animal. And scavengers from long distances die from eating carcasses deliberately laced with poisons, which have killed many undertakers at one blow. Today, as I will indicate, even more potent poisons meant to kill crop-eating rodents kill other animals unintentionally as well.

As I've stated, the most potent human intervention that affects the undertakers is our wholesale removal of large wild animals from the ecosystem and their replacement either with domestic herds for our consumption or with agriculture that removes habitat. But other factors contributing to the destruction of the large undertakers include our methods of animal husbandry, dangerous chemicals, methods of carcass and meat disposal, and cultural practices. Some of these are difficult to rectify, but others could be solved simply by overcoming cultural taboos. One of the most damaging practices affecting the undertakers' livelihood may be our deliberate removal of carcasses that have, throughout evolutionary history, been left to return to the earth. Even parts of animals that used to be "waste" and that could have fed vultures are now processed into hot dogs and the like, although a small concession is made in saving suet for birds such as woodpeckers, nuthatches, and chickadees.

Most parts of any domesticated livestock are now cycled only into human consumption, with scraps converted to pet food. Thus we and our pets are vulture stand-ins. But if an animal that is deemed not suitable as food for us dies, we also deem it unsuitable for availability to others. Even the road-killed deer and other animals that the highway department picks off the roads are disposed of by burying. Vultures would do the job better if we let them.

SINCE THE DECLINE of the large undertakers has been so gradual, hardly anyone notices. A recent exception that brings the disturbing problem to attention is that of the white-rumped vulture, *Gyps bengalensis*. This species nested in trees in colonies near humans and used to be prominent in the morning skies (after the air warmed and the birds could soar) above cities in India. It fed on human remains and helped dispose of other animal carcasses. It was said that these common vultures, coming together in swarms, could dispose of a cow in twenty minutes. These "ecosystem services" were also city services, much like the services ravens performed in medieval European cities such as London.

The white-rumped vulture declined in Southeast Asia over the twentieth century because, as a result of the collapse of wild ungulate populations, fewer carcasses were available. Yet it was still described as "possibly the most abundant large bird of prey of the world." It had ranged through India, Pakistan, Nepal, Cambodia, Myanmar, Bhutan, Thailand, Laos, Vietnam, Afghanistan, Iran,

China, Malaysia, and Bangladesh. Then, during the 1990s, the population experienced a sudden collapse to fewer than 10,000 individuals. The only extant viable populations now are in Myanmar and Cambodia, and this vulture is classed as "critically endangered."

The white-rumped vulture's decline grabbed attention because this bird was common and well known before its very rapid collapse, which was traced to its feeding on livestock that had been treated with the anti-inflammatory drug diclofenac. In vultures the drug led to kidney failure. It does not require many cows to be treated with this medicine to have lethal effects on a vulture population. Modeling now suggests that the observed population declines can result if only one in 760 carcasses contains the drug. And of course nobody thought it necessary to test whether a drug made in America that makes cows well would make vultures in Iran or China or India sick.

Although the most attention has been focused on this species because it was so common, the same drug, which is used worldwide, caused the same catastrophic decline in at least three other vulture species: the Indian vulture, *Gyps indicus;* the slender-billed vulture, *Gyps tenuirostris;* and the Asian king (or red-headed) vulture, *Sarcogyps calvus.*

The European counterpart of the white-rumped vulture, the griffon vulture, *Gyps fulvus,* was once widely distributed over large parts of Europe. This species would undoubtedly have suffered the same collapse — if it were still around to collapse. It had disappeared from Germany by the eighteenth century, due to the unavailability of carcasses. Tiny, isolated colonies of griffon vultures exist now mainly due to reintroductions from captive breeding and "vulture restaurants," where uncontaminated meat is deliberately left out for birds released after being raised in captivity.

The griffon and three other vulture species used to be common in Israel, where they nested in colonies of hundreds. Recently the populations have experienced a precipitous decline from the use of thallium sulfate as a rodenticide.

There is no letup in these assaults, with "newer-generation" rat poisons, such as Havoc, Talon, Ratimus, Maki, Contrac, and d-Con Mouse Prufe II, continuing to come into the market. They contain anticoagulant compounds, and the rodents who eat them either die outright or become weak and thus are easy prey for all sorts of animals. It takes months for the chemicals to be eliminated from the bodies of the predators/undertakers; these poisons have recently been detected in 70 percent of barn owls examined in western Canada, where they are "threatened"; in eastern Canada they are "endangered." But this is only the barely visible tip of a large iceberg: rodents are a favorite snack for countless species all over the world.

Biocides should be outlawed. Rodents can indeed multiply even faster than the proverbial rabbits, as I observed when I left grain for my unguarded chickens. But there are other methods of rodent control. I once easily captured a dozen rats in one night using a drop-in trap made from a trash can; I killed them with a stick. A slower but surer method would be to foster populations of native owls and kestrels. One major limiting factor for these birds is the lack of hollow trees to nest in, but putting up suitable nest boxes in proper locations would help them greatly where old trees are not available.

The New World vulture populations have a mixed record of survival so far. The turkey vulture, *Cathartes aura*, and the black vulture, *Coragyps atratus*, are both doing well, and indeed both are greatly expanding their range northward as carcasses stay unfrozen farther north. The black vulture ranges from southern Argentina to Latin America but has recently expanded into the

Gulf states and most of the Southwest, where it now roosts by the hundreds at night in towns and cities. The turkey vulture, once a strictly southern bird, now breeds even in Canada. For decades I never saw a turkey vulture, but now I see them regularly every summer in Vermont and Maine.

The common raven, *Corvus corax,* like other American undertakers, experienced a vast reduction in its range and numbers after the extirpation of its food base, the carcasses of bison and elk, which had been available over vast stretches of the landscape. Ravens were ancillary kill at poisoned carcasses in the U.S. government's war against predators such as wolves and coyotes. These birds were widely seen as undesirable characters, automatically on everyone's blacklist for extermination, so their deaths by poisoning were deemed irrelevant. Ravens, however, had an advantage that vultures lack; part of their population had hung on in the far north, which still had large animal herds and few people: the north provided a nucleus for regeneration. Poems highlighting old European prejudices against the raven as sinister and dour notwithstanding, the species is making a comeback as a respectable citizen, a valued co-occupant of our earth, and often a close neighbor. The ravens' plight was dismal, but dawning awareness of these birds as intelligent beings endowed with emotions, vitality, and beauty has put an end to the mindless savagery.

TRADITIONALLY, WHEN A great crime such as a murder occurs, people make a great effort to apprehend the guilty party. A murder is a great grievance, but it is a minor one compared to the loss of a species, especially one that is part of a cultural and ecological web that encompasses millions of people, performs ecological services on a near global scale, and enriches the enjoyment of life not just for the living but for all generations to come.

Very few people intend evil; we always have an excuse for our actions. If a death is an unintended or unanticipated consequence, it is categorized as an accident. But accidents are seldom random. It may be an accident when someone drives drunk or goes through a red light in an urgent hurry and kills a person as a result, but it is not a blameless act.

We are all guilty of causing extinctions; our standard of living, our massive industrial production, and our sheer numbers absolutely guarantee that we have a cumulative toxic effect on nature.

What applies to people in the aggregate is true especially of their products. At one time, using a few chemicals out of nature's apothecary was enough to supply our needs. Currently about 84,000 chemicals are in commercial use in the United States, and many are exported to other parts of the world. We don't have any idea what even 20 percent of them are or if they are potentially harmful, because they (and their effects) are classified as "trade secrets."

Of the many people who should be held accountable for the population collapse of the vulture crowd, most are nameless but not blameless. Yet some of us played a more central role than others, and neither ignorance nor personal justification can be a defense when it comes to heinous crimes leading to genocide, ecocide, or the extinction of species. Effects really matter, and any chemical that is synthesized — meaning that it has never before existed as a component of an ecosystem — should be assumed harmful on an ecosystem level until proven otherwise. This is not wild extrapolation; it is commonsense biology. And by common sense it is also obvious that "the market" as such will not solve the problem but will create and maintain it, if left to run free like a high-tech car without a steering wheel or brakes.

III
PLANT UNDERTAKERS

Plants aren't undertakers, but they are the ultimate bio-chemists. Except for several minor exceptions (such as Venus flytraps), they do not consume gobs of flesh or even complex organic molecules. They use water, sunshine, and a few minerals to build themselves out of carbon from carbon dioxide pulled out of the air. However, what they make from these simple beginnings can, by animal standards, become extraordinarily massive and nutritious.

Plants are the intermediary agents in our own recycling into and back out of the soil, and we cannot understand our recycling unless we consider theirs. They are highly adapted organisms whose lives develop according to op-

portunities and constraints similar to those that apply to animals. Like us, they reproduce, grow, and have developed genetic codes inscribed by natural selection on deoxyribonucleic acid (DNA). I concentrate here on the recycling of trees, because they are the most visible plants and overall are perhaps also the most central to the recycling process.

The undertaking of trees (and other plants) is so commonplace in nature that it is easy to take for granted; I admit I used to pay little attention to it. Many animals injure and kill plants, and in any ecosystem where plants live they also eventually die. This process may not be dramatic, unlike an animal carrying off another and tearing it apart in minutes. The death of a tree neither spills blood nor smells bad. Instead, trees may be nibbled on for years by insects, and after they die, through the agency of beetles, fungi, and bacteria, they slowly and unobtrusively disintegrate into the soil, a recycling that makes the rest of life possible and is life as well. The process occurs on a vast scale; were it not for the tree undertakers, the forest would in a few years become an impenetrable tangle of dead wood, soon stopping all plant growth. I hadn't seen much normal tree undertak-

ing in my forest, because most of the tree carcasses that I and others created were cut up and hauled away to be made into lumber, paper, or firewood. In a natural eco-system, however, the dead trees would have been left in place.

TREES OF LIFE

LIKE THE BODIES OF ANIMALS, TREES ARE PREFERENTIALLY eaten while they are still fresh. And since trees have formidable defenses against being eaten when they are alive, they are usually taken only when they are weakest or about to die. The most nutritious part of a tree, the inner bark, which is attacked first, is protected by the tough outer bark shield. But once a tree has fallen, the inner bark is usually available to be eaten for several months, while some of the wood — the framework built to raise the living part of the tree up to the light — may persist for decades.

With the exception (to be discussed later) of possibly one fungus, the world's largest organisms are trees, and some of them are also the oldest, a testament to their ability to resist death by the ravages of parasites and predators. By our own — animal — standards, some trees seem to live forever and may thus appear to be almost inanimate. Every species has its own maximum life span. Some, like the bristlecone pines, redwoods, and sequoias of western North America, may live several thousand years. Some of those individuals living now were already giants at the time of Christ, and they may indeed seem immortal. Most oaks can live several hundred years. The oldest white pines, red spruce, cedars, and sugar maples in my forest can live to about two hundred years. The balsam firs and gray birch may not live to fifty, and striped maples seldom make it beyond twenty. These maximum life spans, however, have almost no bearing on the actual life span of individual trees; most die young.

The trees that we see in a forest are a small proportion of the whole; they are only the survivors and the recently dead. A healthy *forest* (as opposed to a plantation) has dead trees standing and lying about. But the majority of its trees died and were recycled while they were so small that we didn't notice them. Most of my forest has dozens of trees per square foot, but few will produce more than two leaves before they die because of shading or crowding, which amount to the same thing. For the most part, trees in a forest live or die not, as often supposed, by some genetic advantage but simply by the good or bad luck of the location where they became rooted relative to others, which determines whether they win or lose in the competition for light and other necessary resources.

The fate of soon-to-be-dead trees that reach maturity usually begins with insects, and, as with their animal-system counterparts, the "predators" are often not clearly differentiated from the "undertakers" or scavengers. However, a tree is disposed of by a crew or guild of organisms at least as diverse as those that dispose of a mouse, a moose, or an elephant. As in the animal analogue, some of the actors in the tree-undertaking process have evolved to take parts of the body while it is still healthy. Others have a chance only when the tree is weakened, and the majority have to wait until it is almost or fully dead, or even long dead. Scavengers facilitate the latter stages of disposal and bring about transitions. As with animal carcasses, a progression of scavengers attacks the carcass, one species after another, until the feeding queue ends and the tree has returned to soil.

• • •

I WANTED TO SEE how quickly the attack on a downed tree and the subsequent recycling would take place. While building my log cabin in Maine, I axed down about sixty balsam fir, spruce, and pine trees, and I sometimes saw sawyer beetles flying in to lay their eggs on the logs even as I was chopping the limbs off. The sawyer beetles, commonly known as longhorns because of their long antennae, belong to the Cerambycidae family. The antennae of a male sawyer are about twice its body length. They are the insects' chemical detectors, and their great length attests to their importance in detecting the scent of the beetles' specific oviposition sites and of potential mates. The longhorns are so good at finding fresh dead trees that I had to peel the bark off every log I cut. Otherwise they would have attacked the logs by the hundreds and rendered them useless to me.

Besides the longhorns, the jewel beetles (Buprestidae) and bark beetles (Scolytidae) lay eggs on or in bark; the larvae burrow into the cambium of the inner bark and then into the sapwood. As they bore in, they inoculate the tree with fungi, which starts the process of digesting wood, much the way bacteria hasten the decay of animal carcasses. I myself can smell conifers, and surely they can as well, but I never saw a pine sawyer beetle on a healthy vertical pine tree. So how, I wondered, were they able to home in on one I had just chopped down?

I forgot about that question for years because I was studying the foraging behavior of ravens, not beetles. But it came back to me when, in the course of thinking about plant undertakers, I was also thinning out my sugar maple grove to give the trees more space. This time I deliberately left some white pine carcasses on the ground in the woods, and as I'll show, some of these carcasses were not, to my surprise, visited by beetles for months. Why, or why not, and how were the beetles attracted?

It seemed impossible to me that the beetles would sometimes

arrive on the scene so quickly, since a healthy tree that is chopped or sawed down is really not properly dead as far as its tissues are concerned. It is merely condemned to be dead. The beetles that attack it must be banking on its *future* vulnerability, after the larvae hatch. On the other hand, I knew that jewel beetles are attracted to forest fires from perhaps hundreds of miles away, probably to get a jump on the feeding frenzy on the just-killed trees.

Most of these beetles cannot "hunt" in the sense of finding and then killing trees, since healthy trees effectively defend themselves. In the conifers, most famously, that defense is the exuding of sticky, turpentinic resins, similar to the defenses employed by some insects, such as stinkbugs, and even skunks. These defenses are a byproduct of the ancient arms race between tree and beetle, a race that has also spurred an intense war among beetle species. Specialization is a must for beetles. Beetle recyclers have to go after the helpless or the dead; except for some notable exceptions, they are scavengers.

I decided to be deliberate in recording what I observed about my felled pines. On May 11, 2011, I left the trunks in a clearing by the cabin. The temperature was a balmy 60 degrees F. Expecting to see beetles flying in at any second, I waited and waited and waited — for over a month. And still I saw no pine sawyers (the most common longhorn on pine) on any of the logs. I doubted they were extinct. Far from it, as I soon learned.

I woke up in my cabin in the middle of the hot night of July 23, startled by a large insect walking on my naked back. I jumped up and caught the first pine sawyer I had seen that year. The next night another one disturbed my slumbers. The following morning I saw one on the inside of the window, and still another came crawling up my pant leg when I sat down. I had kept the cabin door closed and the screens on, which had effectively held off the black flies and mosquitoes — and presumably the 32-millimeter-long, 8-millimeter-wide beetles as well; the beetle source had to be inside the cabin.

It was then that I thought of the pine log, one foot high and wide, that I had been using as a chair. The year before, I had cut the log and peeled off its bark from a live white pine that had been felled by a spring storm. Inspecting it now, I found some rather large, perfectly round holes — about 8 millimeters in diameter — in its side. I counted nine holes and knew that none had been there in the previous weeks; they were most likely the exit holes of the beetles I had just caught. I sawed my chair into "cookies" and discovered tunnels going clear through the center of the foot-thick pine log. Mature longhorn larvae, pupae, and adult beetles were scattered throughout the wood, deep inside the log. The adults, however, had tunneled almost to the surface; to emerge into daylight they would have had to chew through less than a centimeter more. Apparently at this latitude, late July was the time when the beetles completed their development, which had started the summer before. This explained why I had not seen beetles arrive at my freshly cut pine trees during the last two months of spring and early summer.

As I had predicted, from that time onward and into mid-September, pine borers started to appear at the logs I had put out during the spring, and by early August I could hear the grubs "sawing." This sound, which might be likened to scraping, varies in frequency with the temperature — the pitch is much higher on warm days — and I heard it day and night. The product of this "sawing" appeared to be 1- to 5-millimeter-long wood fibers or splinters, known as "frass," which accumulates on the ground in conical piles below the holes the larvae chew through the bark and deep into the wood. Sometimes the frass almost spewed from the holes, as if the logs were leaking their innards. The frass could not have passed through the beetles' guts: I examined the gut contents of

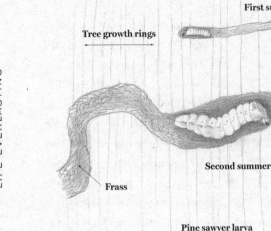

First summer

Tree growth rings

Entrance
into the wood
(bark is here
removed)

Second summer

Frass

Frass

Pine sawyer larva

The view of a cross-section of log showing a pine sawyer larva entering the wood during its first summer (top), and in its expanded burrow in the second summer (bottom).

both adults and larvae and found no trace of this material. The adults had empty guts, and *their* tunnels, just before they exited in late July, contained only a fine, powdery, sawdustlike material. The larval guts contained a smooth creamy paste. Apparently the frass is a byproduct of the larvae's chewing through the wood and perhaps eating part of it, the way we eat nuts and discard the shells.

A month later, in early September, I used my chain saw to gain a view into the pine logs to trace the progress of the grubs. In early August the first larvae that hatched from the eggs deposited by the adult beetles in midsummer had begun making their feeding burrows at the interface of the inner bark and the sapwood. Now, however, there were no larvae under the bark; they had all burrowed deep into the logs. The larvae, which I had often seen in

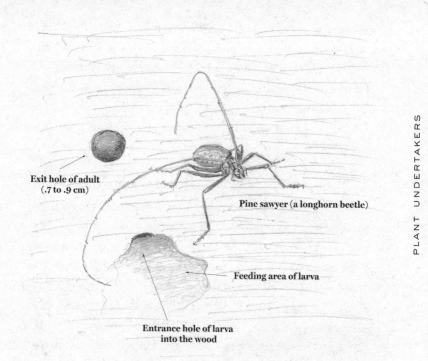

Exit hole of adult
(.7 to .9 cm)

Pine sawyer (a longhorn beetle)

Feeding area of larva

Entrance hole of larva
into the wood

Adult pine sawyer beetle after emerging from the tree, next to the exit hole it has chewed to emerge. For scale is the larva's entrance hole from the previous summer, where it entered the wood after feeding under the bark.

logs during the winter, would pupate there the next spring and then metamorphose to adult beetles and emerge in July or August. I hadn't answered my question regarding how beetles arrived so quickly at downed trees, but I had found out why they were (usually) not around until late summer.

Temperature is perhaps the main variable affecting beetle emergence. Beetles can emerge even in mid-winter, I found out to my surprise one February 1st. Outside temperatures had been at or below 0°F, and on this day I had heated my cabin from its normal 30–50°F to a balmy 75°F. Within hours many beetles started arriving at one of my upstairs windows. Within a day I had collected 353 of them! They were all bark beetles of a species that to my naked eye looked like black specks. None were longer than 2 millimeters or wider than 0.4 millimeters. Their source was the legs of a table I had made in the fall, of a dying white birch tree. I had left the bark on. The volume of the 353 beetles came to one level teaspoon.

ALMOST EVERY SPECIES of wood-boring beetle leaves behind distinctive "tracks" as it burrows through the wood, and each species uses particular species of trees. I went to one of the "healthiest" forests I could think of, the William O. Douglas Wilderness, adjacent to Mount Rainier, not far from the Pacific coast, and hiked along the Yakima Indian Trail under never-logged giant cedars and Douglas firs. I saw living firs of all ages and dead ones in all stages of moldering back into the soil. I peeled bark from one of the recently fallen giants and found hardly a square inch that was not beautifully patterned by beetle tracks — burrows inscribed on the inner bark with a corresponding image on the underside of the wood. They were like the tracks left by the pine borer larvae in Maine, though most of these in the Douglas firs were made not by long-horned beetles but primarily by bark beetles (family Scolytidae), which are generally tiny and inconspicuous but potentially devastatingly destructive. I find many of these beetles in my woods in Maine in trees that are about to die.

The tracks of bark beetles and their larvae make beautiful tattoolike patterns on the wood surface beneath the bark. A recently felled American white ash tree in my woods showed prominent, almost straight lines etched across the grain of the wood — the lines would be horizontal in a standing tree. Each line was an inch or two long. Smaller grooves, largely aligned with the grain, abutted the larger groove at right angles on both sides. Forty to sixty of these vertical tunnels, each one excavated by one larva, radiated from both sides of the central horizontal line. A piece of bark I peeled off a dead balsam fir only a few steps from the ash was being processed by a different species of bark beetle. This one, instead of leaving one main horizontal line, leaves a repeating pattern resembling a brittle star. As in the feeding patterns on ash, numerous small grooves radiated from both sides of each arm. The question naturally arises: How do such strange and "artistic" feeding patterns occur, and why does one species' pattern differ from another?

Almost all of the feeding patterns (except for the short exit tunnels) of longhorn beetles in wood are made by the larvae, but those made by scolytid bark beetles include a significant contribution by the adults to their larvae. Feeding begins when a lone adult male burrows through the bark on a just-dead tree or a sick one that cannot mount a sufficient defense. Each beetle, after reaching the outer sapwood, makes a small cavity below his point of entry through the bark. Then one or several females, depending on the species, come in to join him in this "nuptial chamber." After mating, each female excavates a gallery or tunnel radiating out from the mating chamber below the entry hole. In the above example of the white ash, the horizontal line is actually two adjoining galleries; the balsam fir usually has four radiating galleries, although I saw up to seven, each made by a different female in the male's "harem." Each female deposits eggs right and left at intervals down her gallery, and the larvae that hatch from them make their own, smaller galleries at right angles to their mother's. The number of galleries reveals the number of that female's offspring, and the length of each side gallery is an indication of the amount of sapwood the larva ate before it pupated at the end. After about a month (depending on the temperature) the new beetles leave the tree through the hole that the male made on entry. (The fungi and bacteria introduced by the beetles now speed up the tree's under-

LIFE EVERLASTING

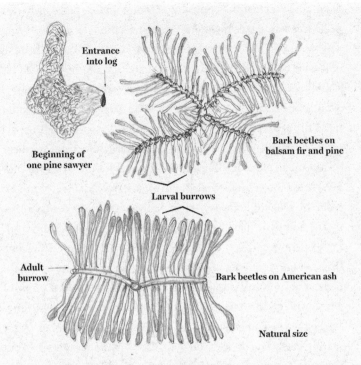

Feeding tracks of bark beetles on logs of balsam fir and pine (top right) and of another species on American ash (bottom). The center burrows are made by the adults, and each radiating burrow by a larva. For comparison, see the beginning feeding track (left) of a long-horned beetle larva on pine.

taking.) The family feeding pattern is thus not artistry but a record of beetle social behavior, sexual pairing modes, and role as tree undertakers.

My peek at the beginning of the recycling of trees at my doorstep highlights how these highly specialized beetle undertakers are in some ways reminiscent of those associated with animal carcasses. It also suggests the important role of temperature in the trees' defenses. Only one generation of pine sawyers is produced each year, and that one cycle takes almost a full year. But very small beetles, such as bark beetles, reach maturity much more quickly. With warm enough temperatures and a long summer, up to six generations of bark beetles can be produced in a single year.

Global warming is permitting bark beetles to produce more generations in a season, and this climate-induced high reproductive rate is causing massive forest destruction in Alaska, northern Canada, and parts of the western United States; the extended warm season is allowing bark beetles to attack en masse and overwhelm the trees' defenses, taking more and more healthy trees that would otherwise be immune to their attacks.

WOOD IS ATTRACTIVE not only to beetles. One group of Hymenoptera, the order to which bees, ants, and wasps belong, are solitary rather than social insects, and their larvae feed on wood. These, the horntails (Siricidae), are large, robust wasps who get their name from the straight, stiff ovipositor sheath at the female's tail end. When ready to lay eggs in a log, she withdraws her needlelike ovipositor from this sheath and points it straight down, at right angles to the sheath and her body. She then drives the hollow ovipositor almost its whole length into the solid wood. If she senses that the wood is suitable, she propels an egg down into the wood, along with fungi and a mucous secretion that promotes fungal growth and helps the larva digest the wood. Like the wood-boring beetle larva, the wasp larva creates a burrow behind itself as it chews through the softening wood.

An insect larva is relatively safe from most predators and parasites deep inside still-solid wood. But one type of ichneumon wasp, *Megarhyssa ichneumon,* has specialized to parasitize horntail larvae. The female of this species has an ovipositor that is up to 10 centimeters long—longer than her entire body (in contrast to the 1-centimeter-long horntail ovipositor). In flight it looks like a long black thread dragging along behind her. This "thread," however, consists not only of the ovipositor but also of two other small threads wrapped around it, forming a protective sheath. Unlike the horntail's ovipositor, this one is very flexible, yet the wasp can drive it several centimeters into solid wood and propel an egg all the way through it and into a horntail wasp larva.

Unlike the horntails, the *Megarhyssa* female cannot use brute force to drive the whiplike, flexible ovipositor. She has to take it out of its protective sheath and loop it over her back in a big bow in order to get the tip of it to even touch the wood, in what looks like an acrobatic maneuver. Her egg-laying task is long and dangerous (while so engaged she is effectively attached to the wood and cannot quickly withdraw, occasionally getting stuck there), so it's unlikely that she would invest in it without somehow knowing where her target is located deep within the wood. How she finds that out is not known.

WHEN THE WOOD-BORING beetles and horntails emerge from a recently dead tree to complete their life cycle, they leave behind a new habitat that is suitable for numerous other insects. The galleries in the wood made by beetle and wasp larvae are used by a variety of insects. First, those beetles that feed on fungi rush in, followed by specialized predators such as *Colydium lineola*, which feed on the first colonizers. One group of bark beetle predators, the clerids, colorfully decorated in patterns of red, orange, white, and black, have thick heads that anchor the powerful mandible muscles needed to crunch their beetle prey. (In contrast, note the tiny head of a pollen-eating beetle.)

Eventually the bark starts to loosen from the tree, creating even more habitat for other insects and spiders, who find food and shelter there. These colonizers then attract their own predators, such as the red flat bark beetle, *Cucujus clavipes*.

As long as wood stays dry, it resists decay, yet the larvae of some specialist beetles do eat dry wood, including various species of tiny brown beetles known as powderpost (Lyctinae; Bostrichidae), and deathwatch or furniture beetles (Anobiidae). The larvae process the wood for the small amounts of nutritious starch in it; their burrows, which may be only 1 to 3 millimeters wide, spill powdery sawdust as they chew. Moisture can now enter the wood through these pores, allowing decay to proceed more rapidly.

Increasingly softened by burrows and then fungal and bacterial decay, the wood eventually becomes suitable for the larvae of the large spike-edge long-horned beetles. Meanwhile, the fungal undertakers produce fruiting bodies on which certain beetles feed.

In the tropics, moist decaying wood and other plant matter become the habitat of scarabs, including the world's largest beetles, the South American Hercules beetle, *Dynastes*, and the African Goliath beetle, *Goliathus giganteus*. The Goliath lives up to its name; it may be up to 10 centimeters long, and the larvae weigh up to 120 grams, about ten times the weight of a warbler. Here in the northeastern forest, the only wood-eating scarab beetle I am familiar with is the all-black *Osmoderma scabra*, whose fat white larvae I find routinely in almost any hardwood tree that has damp decaying wood. This larva is partially transparent, so you can see the dark wood mush in its digestive tract. Farther south one would also find the larvae of the mostly tropical fruit and flower chafers in a subfamily of scarabs, the Cetoniinae, of which the Goliath beetle is a member. Worldwide, there are an estimated four thousand cetoniine species. Most are tropical and many of them are so far undescribed.

The cetoniines' size range is enormous, as is the spectacular variety of their bold, bright, usually metallic markings. Although all the larvae are white and live on decaying vegetation, the largest ones primarily eat rotting wood, while the adults feed on decaying

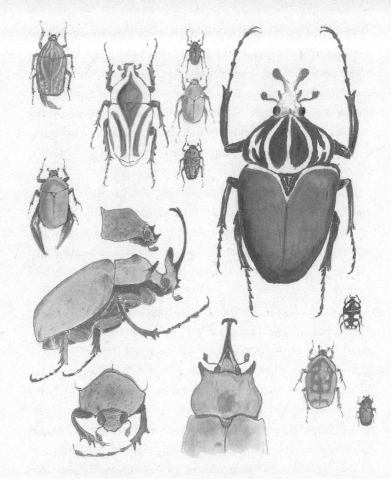

Fruit and flower scarabs. All but one of these cetoniines are from East Africa;
the South American elephant beetle, megasoma alphas *(lower left), is shown*
in four views, male and female. The two giant scarabs, the rhinoceros and the
African Goliath beetle, Goliath giganteus *(top right), are fruit eaters as adults,*
while the others are pollinators mostly of trees. All of the larvae feed on decay-
ing wood and other dead vegetation. Flower scarabs are renowned for their
brilliant colors. These are resplendent in metallic greens, yellows, and rich
browns.

fruit. The adults of intermediate-sized beetles of this group eat
flower petals, and the smallest ones eat pollen.

Because of the adults' feeding patterns, the cetoniines are major tropical pollinators; the flowers of many plants are specifically adapted to be pollinated by them. In a recent study of two orchid species in South Africa that offer no food rewards (no nectar or pollen is available to feed visitors to the flowers), it was discovered that the plants did not set fruit if cetoniine beetles did not visit. The beetles apparently visited (and pollinated) the orchid flowers because they mimicked those of a plant that *did* offer food. Thus both plant species must exist in the habitat at the same time in order for the beetles and the plants to live there. I recall with pleasure seeing and hearing flower scarabs zooming around flowering acacia trees on the southern African savanna. Near Dar es Salaam they were flying around, and presumably pollinating, blooming mango trees, and I found their large white grubs in rotting tree trunks on the ground. Thus, in the continuing processes of death turning to life, some of the beetles that help in tree undertaking also play a crucial role as surrogate reproductive organs in a direct interactive living system. These interrelationships exist in all biological communities but are seldom so direct and simple.

ONCE SEEDED INTO the dead tree, fungi account for most of the decomposition of wood. Indeed, according to the mycologist Paul Stamets, fungi can "save the world." Their role in our lives includes the provision of food and antibiotics and the neutralization (as well as production) of toxins. But I suspect that all of these services dwarf their role in decomposing wood, which helps build soil.

A fungus assumes many forms. Sometimes you see it, but most of the time you don't. After extracting sufficient nutrients from a tree, the fungus converts them for its own reproduction, growing what may be a highly visible and even spectacular fruiting body.

These fungal reproductive organs, which produce and spread spores, are usually known as mushrooms, conks, or brackets. The main body of the fungus that produces these structures is a thread-like net growth called, in aggregate, the mycelium, which can grow for many years in a tree trunk before producing fruiting bodies in response to particular conditions of temperature and moisture. Tubes or lamellae on the ventral surfaces of the fruiting bodies then release millions of spores, which travel mostly by the wind. After landing on a suitable place, a spore germinates and produces a new mycelial web. Two mycelia of opposite mating types may meet and join to produce sexual spores.

Most mushrooms last only a few days before decaying or being eaten, often by gnat (fly) larvae. However, some mushrooms — conks — last for many years, each year adding a new spore-bearing layer at the bottom. The age of other fungi, such as some that grow on soil, can be measured by their periodic production of fruiting bodies. On the lawn of a neighbor of ours, a flush of mushrooms comes up some years, always in a circle, which year by year becomes larger. After a week or so the mushrooms are rotted and gone, but the mycelial body that produces them remains underground, to send up its fruiting bodies in subsequent summers.

Fungi living on trees are similarly hidden most of the time. Take the maple-decaying fungus *Armillaria mellea,* which forms a white fungal mat under the bark of the tree. This fungus is bioluminescent — it glows in the dark — but of course you cannot normally see this from the outside. I have observed only the later stage of the fungus, when the bark is dead and loosened. You then see a dense network of black "shoestring" fungal organs called rhizomorphs (root forms), which are visible for months or years. A third form of *A. mellea* is its fruiting bodies, little brown mushrooms that produce the reproductive spores. These honey mushrooms come up at the base of the infected tree, shedding their spores and deteriorating in only a week.

Edible mushrooms are a gourmand's delight, and when my family was living in the Hahnheide forest, we indulged. We were scavengers living on scavengers, mainly the ones we called *Rehfusschen* and *Steinpilzen*, along with many others whose names I no longer remember. One mushroom that is currently a big hit in the United States is *Lentinula edodes*, which has been cultivated in Asia for thousands of years. It is generally known by its Japanese name, shiitake (from *shi*, meaning "oak," the tree on which it grows). Shiitake mushrooms are prized for their taste and for their reputed enhancement of immune function and high protein content. They are cultivated on freshly cut — but not too fresh — logs and are currently being grown on oak and maple logs even in my neighborhoods in Vermont and Maine. Shiitake "spawn" is commercially available, and I plan on using it to recycle the sugar maple logs that I must weed out of my maple grove. Instead of waiting for beetle larvae to inject the fungal spawn into their logs, growers make cuts with a chain saw, rub the spawn in, and then seal each inoculum in with melted wax.

Here in New England, several other mushrooms that live on freshly dead or dying hardwood trees, especially oaks, are sought as food by many people. We scour the woods every late summer and fall for the sulfur shelf (*Laetiporus sulphureus*), also called chicken-of-the-woods, because it tastes like ... chicken. The fruiting bodies produced by one fungus in a flush of "fruiting" may weigh more than fifty pounds. Another mushroom, the hen-of-the-woods (*Grifola frondosa*), whose fruiting bodies are almost as big, is delicious, as is a third disposer of wood, the oyster mushroom (*Pleurotus ostreatus*), which grows on dead deciduous trees, especially beech. The fruiting bodies of these fungi may tickle our palate and serve as a vital food resource for many

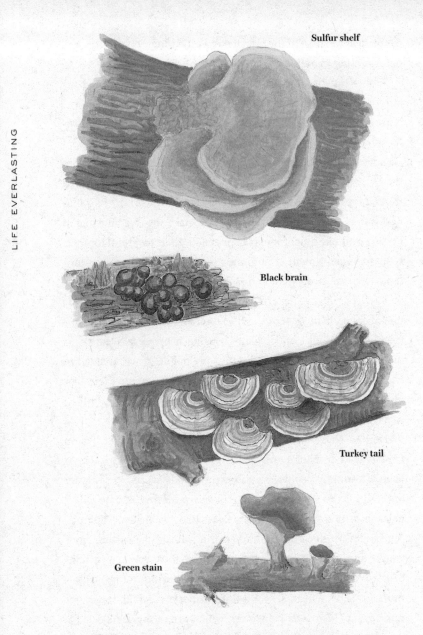

Sulfur shelf

Black brain

Turkey tail

Green stain

Fruiting bodies of some of the many fungi that recycle dead wood. Colors range from brilliant red, yellow, and green to black and brown.

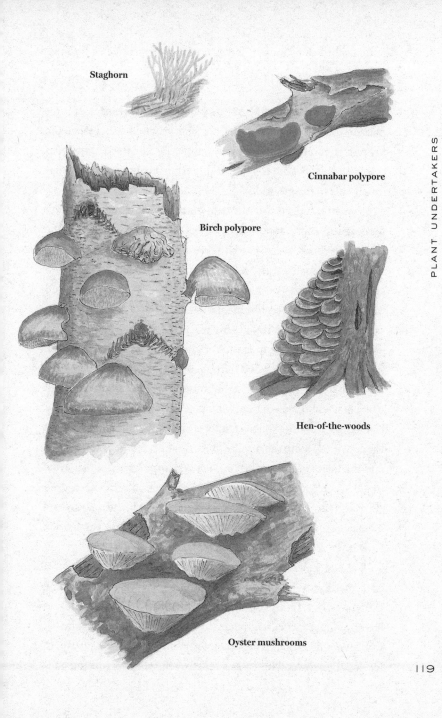

Staghorn

Cinnabar polypore

Birch polypore

Hen-of-the-woods

Oyster mushrooms

animals, but their unseen forms do far more service as tree undertakers.

TREE UNDERTAKING PROCEEDS at a glacially slow pace in comparison to that of animals, but on their way to death, before the process is complete, many trees supply life. In this transitional phase, the body of the tree serves important ecofunctions even before it falls to the ground.

Even after a dead tree starts to decay, it may remain standing for decades. These standing dead trees are a prime indicator of the health of a forest because, well — because they are an indicator of its life. More than a third of the bird species in a forest depend on standing dead trees, both for their food if they eat beetle grubs and for nesting places, because the partial decay makes nest-hole construction possible. Without this interim stage, most woodpeckers would not be able to exist. Very few can hammer a nest hole solely out of solid live wood (although they may hammer through an outer solid layer to get to the softer, partially fungus-softened wood beneath it, where they will make their main nest cavity).

One of the more explicit examples of this phenomenon involves the tinder polypore (also called hoof fungus), *Fomes fomentarius*, and the false tinder mushroom, *Phellinus igniarius*, which grow in old aspen (poplar) trees. The tinder mushroom has been used since ancient times to start fires from sparks (Ötzi, the 5,300-year-old Iceman discovered in 1991 in an Italian glacier, carried it). Now it serves mainly sapsuckers.

The late Lawrence Kilham, a physician and ornithologist of note, made a study of the yellow-bellied sapsuckers near his home in Lyme, New Hampshire. He determined that the sapsuckers have an apparent search image for the mature fruiting bodies of the tinder mushroom. The fungus grows in the aspens' heartwood, leaving the sapwood intact as a tough shell, while its fruiting bodies are on the outside, attached to the bark. These fruiting bodies lead the birds to the trees in which they excavate their nest holes. Unlike many other woodpeckers, the sapsucker doesn't excavate wood to get insect larvae. Instead, it makes sap licks by puncturing the bark of maple, birch, basswood, oak, and other trees. Perhaps these woodpeckers prefer softened-up poplar trees for their nest holes because they are unwilling or unable to excavate solid wood.

Before knowing of Kilham's 1971 study, I had confirmed his results; I had wondered if these woodpeckers had a preference for aspen trees with the fungus, because I always saw the fungus whenever I found a sapsucker nest hole. I surveyed the trees in my Vermont neighborhood, where there are many poplars. I examined 176 poplars along our road. Of these, 12 had *Fomes* conks, and 5 of these 12 had sapsucker holes; there were no sapsucker holes in any of the aspens that didn't have the fungus. Sapsuckers seem to deliberately choose as their nest tree one that has a soft core. Perhaps they identify it by the fruiting bodies of the fungus. The other local woodpeckers also use poplars, but they do not have a preference for them or for trees with the fungus. Downy and hairy woodpeckers appear to choose a high, still-solid dead maple stub for excavating a nest hole in which to raise their young. However, in the fall they often excavate a much more rotten, lower stub as an overnighting cavity. Two holes made by hairy woodpeckers that I recently found were in a long-dead balsam fir and in a birch softened by the parchment fungus, *Stereum rugosum.* Two downy winter shelters were made about 2 meters above the ground in sugar maples taken by the turkey-tail fungus, *Trametes versicolor.*

Despite these preferences, if they are available, woodpeckers are flexible, though not necessarily with the best results. My cabin site in Maine has few poplars, and I once found a sapsucker pair who had made their nest hole in a dead maple stub. I made this discovery inadvertently after the stub broke off in a storm, spill

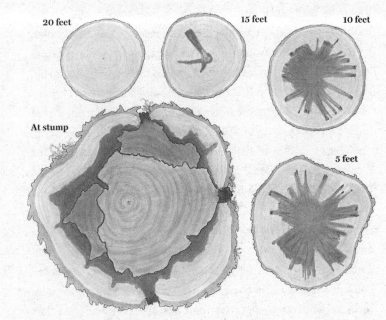

Progression of a fungus in a live sugar maple tree that was being shaded and would likely soon die. The fungus probably entered near the base (lower left) where live tissue had grown around a physical injury, leaving three exterior wounds. The lighter heartwood is dead, rotting (though still solid) tissue; the black areas are where the tree is fighting the infection. The cross-sections show that the fungus has extended up the tree to a little over fifteen feet.

ing the still-naked babies onto the ground, where I found them dead.

The babies of all the woodpecker species I know are very noisy, making a nearly constant raspy din. Possibly the noise helps motivate their parents to continually feed them. It must also attract predators, although for the most part the young woodpeckers remain safe inside their fortress, the solid tree. Kilham found, however, that raccoons could sometimes extract young sapsuckers from their nest, provided they could break into the nest hole.

Raccoons had little success reaching the noisy sapsucker young in aspens, which have the hard shell of sapwood on the outside, but they could break into nests when the woodpeckers chose dead maples, birches, or beeches — trees that lack the tough sapwood shell.

An aspen infected with false tinder fungus can be a valuable resource for yellow-bellied sapsuckers. Once they find one, they may return to it for several years in succession to nest; sapsuckers are the only woodpecker species that returns to the same tree (but, like other woodpeckers, it makes a new nest hole each time). A sort of "tenement house" results if the birds return for six or seven years. Some of the empty holes are recycled as homes by northern flying squirrels and also by nuthatches, titmice, and chickadees.

Hairy, downy, and pileated woodpeckers preferentially choose hardwood trees for their nest holes, using parts of the tree that are dead but still solid (at least on the outside), and usually very high up in the tree. Unlike the sapsuckers, these woodpeckers do not migrate south in the winter. The hairy and downy woodpeckers are the only birds I know in New England who build themselves a shelter in which to stay overnight in the winter. In October they excavate cavities similar to nest holes, except that they are almost always in decayed, easy-to-excavate tree stumps. In turn, in and at the edges of forests, three owl species, as well as wood ducks, mergansers, nuthatches, crested flycatchers, tree swallows, bluebirds, and sparrow hawks (American kestrels), nest in one or another kind of woodpecker hole. Black-capped and boreal chickadees sometimes excavate nest holes, but their tiny bills are much weaker than the sapsuckers'; they cannot penetrate solid wood. They seek out wood that is quite decayed. Brown creepers do not excavate nest holes, but they also require dead trees because they build their nests under hanging bark, mainly on dead conifers. In the tropics, almost all parrots, hornbills, barbets, and many fly-catchers depend on holes in trees for their nests.

TREES, ALIVE AS well as dead, are also life-promoting for fish. Trees growing along stream banks shade the water and keep it cool, which helps trout breathe. They need lots of oxygen, and warm water holds little of it. But beyond that, the brook trout, *Salvelinus fontinalis*, a beautiful green-marbled char, with red dots surrounded by blue haloes, red-edged fins, and a pink or red belly, needs places to hide and to rest — as all organisms do. The roots of trees along the banks of a brook play a pivotal role in holding the soil, while the rushing water undercuts the banks, creating cavities where the trout lie in wait to snatch insects drifting by. Once the trees die, they are recycled into the stream.

Along the dirt road where I live in Vermont is a drainage where, were it not for beavers, the water would be seasonal. Thanks to the beavers, who harvest trees for food and for dam construction, the water is now there year-round. The beavers have built a series of dams (fifteen at my last count) that impound the water in steps down the valley incline. The dams range from about twenty to several hundred feet in length. The largest of the beaver-made ponds holds three species of fish and is the breeding place of six species of frogs, one toad species, and at least two types of salamanders. These ponds are too shallow and warm in the summer to be suitable for brook trout, but at higher, cooler elevations, some beaver works are prime trout habitat.

Although beavers haul a lot of wood into the water to make their lodges and dams, streams are also dammed up, creating fish habitat, when trees fall over into the water on their own. Over the years, spring freshets may move a tree downstream, where it lodges against the bank, some boulders, or other trees to make a logjam. Water swirls over, under, or around the jam, gouging holes and making pools where the trout hide and stay cool on hot summer days when water levels are low. These rare bottlenecks make all the difference to trout and salmon survival.

SIMILARLY IN A FOREST, the standing dead trees eventually fall and create an ecosystem of quite another set of organisms. But the tree as ecosystem keeps changing, as the work of the fungi and bacteria progresses. Moisture near the ground, plentiful oxygen, and warmth all help fungi soften the wood. Moss grows over the decaying trunk, holding rain water that would otherwise run off. Seedlings of some trees, such as yellow birch in my forest, which are seldom able to punch through the heavy annual leaf fall on level ground, can get a start on moss-covered fallen logs. From

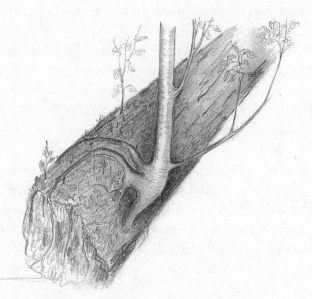

A yellow birch tree growing on an old pine log. Rotting logs serve as nurse trees by providing a space above the leaves covering the ground where tree seeds with minute food stores can get a root hold.

reading the accounts of the Swedish naturalist Peter Kalm, written during his travels in North America in the mid-1750s, I suspect that these moss-coated "nurse trees" are especially important for maintaining some tree species in mature old-growth deciduous forests. While Kalm was in Pennsylvania, he marveled at the large size of the trees and the sparseness of the forest understory. Hordes of squirrels abounded, and people let their pigs range in the woods to forage on nuts. Some hint of why many nut-bearing trees were favored in these forests can be gleaned from his comments on November 13: "The leaves have at present fallen from all the trees, both from the oaks and from all those which have deciduous leaves, and they cover the ground in the woods six inches deep." Only seedlings from large seeds such as nuts can punch up through such a layer. Seedlings of small-seeded species use fallen trees as a competition-free launch pad for their race to reach the light of the canopy.

As a log starts to disintegrate, it begins to admit centipedes and millipedes, and in the fall wasps, beetles, and other insects burrow into it to hibernate. In subsequent years, as it sinks further into the ground and is covered with leaves in the fall, the wood returns to the soil as humus.

ABOUT TWENTY YEARS ago, scientists from the College of Forestry at Oregon State University started a two-hundred-year study of 530 rotting logs in the Cascade Mountains. The study has a long way to go, because huge logs take a long time to decay. But already, as Mark E. Harmon, a professor of forest science at OSU, has said, "Much of what we've found has run contrary to conventional wisdom."

Their main finding so far is that rotting logs feed the nutrient cycle and are far more important to forest health than previously supposed (by some). The availability of nitrogen is a key limiting factor in a forest's growth, and the decay of rotting logs releases nitrogen for reuse. But even more important, the decay process extracts gaseous nitrogen from the air and makes it available for organisms to turn into protein. Another point is that the brown-rot fungi — a group that cannot break down the lignin component of wood — leave behind structural material that helps build soil. The white-rot fungi degrade all parts of the wood, but they act only in some tree species and at varying rates, depending on the species. The species composition of a forest thus has long-term implications for the soil and for regeneration. I suspect it will turn out that tree diversity ultimately enhances tree growth through its effects on the soil and that soil is created not only from the trees' remains at death but also from the shedding of dead leaves throughout their lives.

Every fall the manicured pet grass of suburban lawns receives a blanket of leaves from the surrounding birch, ash, and maple trees. Many people conscientiously rake up the leaves (or, worse, remove them with noisy, gas-powered leaf blowers) as though they were some kind of trash. They stuff the leaves into black plastic bags, which they leave at the curb for the garbage truck to haul off. I leave leaves where they fall, and rain and snow flatten them onto the ground. In early spring, after the first soaking rain and before the grass starts to grow, the leaf undertakers (earthworms) come up out of the ground at night to begin their work. They stretch up out of their burrows, grab the limp, wet leaves in their mouths, and pull them toward and into their tunnels. In the morning you may see leaf tufts standing upright all over the lawn where worms were halfway done with a leaf at daylight. Most worms pause their work at first light and retreat underground; those who don't risk the robin. The more leaves the worms find, the more they multiply, to fertilize and aerate the lawn and make the grass grow.

Soil making in forests is similar. The Maine woods that I am familiar with are wild and wonderful despite logging, and what makes them appealing and keeps them wild is their hugeness, which largely discourages "cleanliness" and encourages litter and slash. There is time for decay and regeneration of the trees back into humus and soil.

Forest soil is a complex, species-rich ecosystem that in some ways acts like an organism itself. Edward O. Wilson described (in *Biophilia*) a handful of soil thus: "This unprepossessing lump contains more order and richness of structure, and particularity of history, than the entire surface of all the other [lifeless] planets combined. It is a miniature wilderness that can take almost forever to explore." Here I will explore it only briefly. Bacteria in forest soil take the ammonium that results from the decay of protein and turn it into usable nitrate. Other bacteria fix gaseous nitrogen from the atmosphere and put it back into the soil. The kinds of bacteria that thrive in and depend on the soil in turn determine the kinds of plants that can grow there. In oxygen-poor environments, other bacteria act as denitrifiers: they return nitrogen to the atmosphere. Another group, the actinobacteria, decompose organic matter to form humus, as do a huge variety of fungi. Still other fungi live in a symbiotic relationship, known as a mycorrhiza, with the roots of trees and other plants. Mycorrhizae are necessary for the tree to absorb nutrients from the soil.

By decomposing dead plant and animal matter, soil microbes release organically bound nitrogen and phosphorus in forms that the plants can use for growth. Thus, over the long run, the forest soil needs dead trees, or the slash and "waste" from logging, to feed it. Aside from the complex chemistry, however, soil with organic matter folded into it has a texture that binds water, making it continuously available for the trees' growth. The carbon, nitrogen, and water cycles all meet in the soil, intersecting on dead trees, which give the forest life.

Soil plays a central role in the productivity of forests, and hence it gives farmland derived from forests their fertility. Soil is currently an especially hot topic not only because of forestry and agriculture, but also because of its role in atmospheric carbon dioxide and climate change. The carbon captured by trees may be stored in the wood of trunks and deep roots and not be released for centuries, or if sequestered into leaves that are shed and incorporated into soil, it is released back to the atmosphere in a year or two. At any one time, soil is about 60 percent carbon. How does the soil carbon affect atmospheric carbon dioxide, which has increased from about 275 ppm (parts per million) at the beginning of the Industrial Revolution to 389 ppm now? The release of carbon dioxide from the soil is under the control of the soil microbes and fungi, and rising temperature increases their activity. It is estimated that Arctic soils contain around half the world's soil carbon, but it is for now not releasable because the soil is "permanently" frozen. A slight warming of several degrees, however, can melt the permafrost and have a huge effect on atmospheric carbon concentrations. Trees "capture" released carbon, but they can't hold it for long if it doesn't stay in the soil. Recent research shows, surprisingly, that when trees take up more carbon dioxide they also release more from the soil they grow on. Possibly faster tree growth results in nutrient release from the roots that stimulate the soil microbes and fungi that then release the sequestered carbon from the soil.

We may think we are God's gift to earth as the greatest undertakers when we dispose of old or dead trees and plant young, "healthy" ones to catch excess carbon out of the atmosphere. We think we do nature one better by planting preferentially "superior," genetically engineered trees that are supposedly more "green" than those that nature has made over the course of four billion years of selection, in which only one of a tree's tens of thousands of offspring may survive to reproduce. However, insects, molds, bacteria, and beavers have engineered the most amazingly foolproof, the most effective and intricate, cooperative system-solution to the death-into-life cycle of trees. The system has been tested in real-life conditions over eons, and it can hardly be improved on by our tinkering with details designed for our perceived immediate benefits.

DUNG EATERS

What can't be used is trash; what can, a prize
Begotten from the moment as it flies.
— Johann Wolfgang von Goethe, *Faust*

IN THE MID-1970S I TRAVELED TO KENYA'S TSAVO NATIONAL
Park with George A. Bartholomew from the University of California Los Angeles. It was Bart, who had been my PhD adviser, who
inspired me to become a physiological ecologist. We went to Kenya
to study the physiology, behavior, and ecology of dung beetles, with
an initial specific interest in the elephant dung beetle *Heliocopris
dilloni.* This sparrow-sized beetle is built like a tank, flies like a
hawk, and tunnels through hard-packed soil like a bulldozer. Each
monogamous pair rears one offspring on a diet of fresh elephant
dung, which the parents carry down into their underground nest
and then make into a ball.

I had seen my first African elephant dung beetles a decade earlier, during my undergraduate studies at the University of Maine,
when I took a year off to accompany my parents on a bird-col-
lecting expedition to Tanganyika (now Tanzania) for the Peabody
Museum at Yale. One of my activities during that yearlong expedi-

tion was to set up nets in the forest around Mount Meru to catch birds. When I checked these nets early in the dawn, I found many dung beetles in them whenever elephants were near. While flying at night toward fresh dung, the beetles had become entangled in the nets. Four years later at UCLA, I studied the physiology of insect flight and its relation to body size and temperature. Large insects were hot, and small ones were not. Studying elephant dung beetles, which were larger than any I had ever seen, seemed a near "must" to round out my studies. I could not get any of them to fly in the lab, at least for long enough to stabilize their body temperature. But it would be easy to measure the temperature of flying dung beetles by intercepting them where I knew they would arrive: at fresh elephant dung. Many species of beetles, in a variety of body sizes, would come at the same time, and these would be the much-needed controls. It was not hard to convince Bart, a world authority on birds and mammals, that this would be an interesting project, and he graciously financed the trip and came along himself.

As many as 150 species of dung scarabs may be found in any one locality in Africa; 780 species live in southern Africa alone. At Tsavo it was the beginning of the rainy season, peak time for dung beetle activity. We came upon a herd of a hundred or so red dust–coated elephants slowly ambling along, grazing on the green grass that had just sprung up. Chalky white butterflies fluttered on scattered low bushes with white flowers. Metallic green cetoniine flower scarabs were humming around acacia trees ablaze in yellow bloom. Elephants were pulling up tufts of grass with their trunks and deftly stuffing them into their mouths. Each elephant daily processes hundreds of pounds of grass and twigs and at regular intervals leaves droppings the size of basketballs and the shape of yeast rolls.

As the elephant herd travels along, consuming a swath of veg

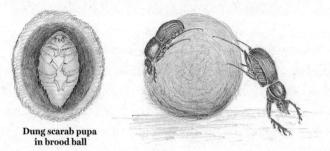

**Dung scarab pupa
in brood ball**

A pair of dung-ball-rolling scarab beetles (right), with the male pushing the ball and the female riding on it passively. The ball, once buried, provides food for their larva, which then pupates inside the ball, as shown in cross-section (left).

etation as they go, they leave a trail of dung piles behind. A casual daytime tourist might be impressed by the number of dung beetles at each pile, but the scale of their activity during the day is as nothing compared to the astonishing spectacle after dark. Bart and I would have our shot of whiskey in the evening along with others at the Tsavo Lodge, but we were here for beetles at elephant dung, not elephants, and we had to be out at night when most of the beetles become active.

If you go out after dark where a herd of elephants has recently come through, or if you bring a bucket of fresh dung collected in the daytime and leave it on the ground, you will hear, shortly before the deep coughing roars of the lions begin, a low humming noise. It's the sound of dung beetles barreling in — hundreds and then thousands of them — all making a beeline to your dung pile. They range in size from slightly larger than a rice grain up to that of the sparrow-sized *Heliocopris dilloni*. We once collected and counted 3,800 beetles coming in to a half-liter sample of fresh elephant dung in fifteen minutes! The total beetle weight exceeded that of the dung we had put out. Every night the dung was eaten on the spot, pulled underground into tunnels, or made into balls and rolled away to be buried elsewhere. After an hour or two, all that remained of the elephant dropping would be a two-meter-diameter pancake of loose, almost dry, fibrous material from which almost all of the juicy goodness had been extracted. But hundreds of rice-grain-sized beetles remained embedded in it a bit longer, trying to extract the last tiny nuggets of nourishment.

THE BEETLES THAT soon interested us the most were the numerous dung-ball rollers of one large-bodied species, later identified as *Scarabaeus laevistriatus*. These beetles came precisely at dusk, just as we were getting started on our night's work. They approached fresh dung in a gingerly walk, palpated it with their antennae, then began to cut into it with the rakelike prongs on both front legs as well as a shovel-like extension on the front of the head. Each beetle pried off some dung, patted it with its front legs, and then pulled additional material from the pile. In this way it proceeded to sculpt an almost perfect sphere the size of a golf ball or as large as a baseball. The process could take from ten minutes to half an hour. When the ball maker was finished, it placed its long, slender hind feet on the ball, almost doing a handstand, with its front feet on the ground. Facing away from where it was going, it started walking backward on its front legs, while its hind feet kicked the ball to roll it along. Before it got a good rolling start, however, hordes of other beetles flew in, planning to steal the ball maker's labor of and for love; some balls made by males are nuptial offerings or sexual displays, and they are prized by competing males.

Newly arriving *S. laevistriatus* beetles often inspected the whole dung pile but showed no immediate interest in ball construction. Instead, they approached a ball maker and jumped onto its almost finished ball. If it was a female and the ball maker a male, the female would flatten herself against the dung ball and stop moving. (In some species the female is the active partner who rolls the ball, and in others both partners roll it.) The male would then accept her and roll his ball away, paying no more attention to the hitchhiker, a mere bump on his ball. But very often two beetles, presumably both males, faced off on the ball and started a sparring match in which each one tried to flick the other off; the ball maker was unwilling to relinquish his creation to another male.

The single roller or couple proceeds in a probably random but consistent direction, possibly by maintaining a constant angle to some landmark or sky-mark. After traveling a suitable distance or arriving at a patch of soft ground, the beetles bury their ball, presumably in much the same way *Nicrophorus* beetles bury a mouse carcass. If it is a couple, the pair excavate a nest chamber and then mate, after which the male leaves. The female lays her one egg and stays on to care for the brood ball, which serves as both crib and larder for the developing larva. (The food may have the same volume as a carcass used by a burying beetle, but it has a lot less protein; hence the single offspring rather than the dozen or so of a burying beetle.) The dung-ball rollers have a life span of up to two years and thus may nest several times in their lives, each time with a different mate.

In the burial process the surface of the ball becomes coated with the dominant clayey soil. It is further reshaped underground and fortified with saliva, creating a rough shell. The female guards and maintains the ball until the larva has eaten away most of the dung and has pupated inside the hollow thus created. The female then leaves. The fresh pupa that remains underground is at first soft and milky white, but one can see the imprints of legs and other body parts, reminiscent of an Egyptian mummy wrapped in cotton gauze. After the rains come again and soften the soil, perhaps a year after the egg was laid, the adult beetle emerges from its "mummy case" by bursting through the shell of dung and saliva and burrowing up out of the earth. It probably waits until nightfall to fly over the veldt in search of a fresh dung odor plume.

I wondered why elephant dung (and many other kinds) is such a prized resource. Isn't dung a waste product? Why would elephants not extract all the nutrients from the vegetation they ingest? The current hypothesis is that the food is shunted through their gut so quickly that they only skim off the cream, so to speak. But if thousands of beetles find sufficient nourishment in a half-liter of dung, it would seem that many nutrients must be left. Do elephants not have strong selective pressure for energy efficiency? That can't be true, since their very large size and energy demands should create a high selective pressure to extract every ounce of energy from everything they eat. I realized, however, that they achieve their efficiency in digestion by relying on symbiotic microorganisms in their gut, which have the enzymes that break down the roughage they ingest. But the corollary is that these magical symbionts are to the elephants what cows are to us. Their "owners," who herd the microorganisms in their gut, must practice restraint. That is, they can't evolve enzymes that would kill their symbionts in order to digest them for a quick profit. They must let them live, and indeed elephants may have some unknown mechanisms that block them from digesting their "cows." Also, these commensals must have evolved mechanisms that keep them from being digested by their hosts. As a result, some of these organisms end up whole in the feces. Once shed, however, they are a potential source of protein for those who have no economic restraint on killing the geese that laid the golden eggs for the elephants. And so I suspect that the dung beetles enjoy a free ride, thanks to the gut symbionts that the elephants need but can't keep completely.

Dung is a valuable resource to many animals, and a system for recycling it probably evolved soon after it became available on the landscape. Cretaceous deposits in the Two Medicine Formation in Montana have shown that dung-processing beetles had already evolved relatively "modern" burrowing behaviors, and there were many such beetles at that time: they dug burrows apparently to remove the dung they collected from the intense competition aboveground. Given the many dung-producing animals — elephants, buffaloes, antelopes, giraffes, warthogs, baboons, lions, hyenas, jackals, leopards, hippopotami, humans, and rhinos — there is plenty of it and lots of variety. As animals drop dung, they, with the aid of the recyclers, produce soil and often spread and plant seeds of plants that, like the microbial symbionts, are resistant to digestion and can be transported alive in the guts of animals. Elephants, for example, are fond of fruit, and they spread great numbers of seeds from the many fruiting plants they eat. Some plants rely entirely on elephants for their seed dispersal, much as others rely entirely on certain species of wasps or bees for the pollination of their flowers and thus their propagation. A recent study has shown that Asian elephants spread seeds over a distance of one to six kilometers; Congo forest elephants spread seeds as far as fifty-seven kilometers.

As I examined my collection of dung beetles from Africa and started to draw them, I was impressed by the perfection of their unique shapes and sizes. I first drew the largest, *Heliocopris dilloni*, which lives exclusively off elephant dung. When these beetles come flying in to an elephant dung pile, they crash-land, then fold their large wings and tuck them under their burgundy-brown wing covers. They lumber along slowly, looking like muscular weight men on a track team. But these miniature bulldozers can tunnel straight down into the ground. For digging they have a four-pronged bulldozer-blade extension on the front of their flattened head, and their front tibia have shovel-like lateral extensions to push loosened soil out to the sides. Their hind legs are short, thick, and packed with muscles, and the ends of their hind tibia act as "treads" to harness all that power. Instead of coming to a single point, as in most beetles, the tibia of *H. dilloni* present a roughly square surface with several backward-pointing spines, to provide traction as the beetle pushes head-first through the ground. After digging a tunnel under the elephant dung pile, the male meets a female there and then stays near the top of the tunnel to bring loose dung down to her. She then makes the dung ball that will serve as her oviposition site and as the food for their larva. In some tunneling species the male brings down enough dung for the female to make several balls, and thus several offspring are produced in one nesting.

In contrast to the bulldozing *H. dilloni* are the rollers. These beetles come in all sizes, but all have the ability to remove dung from where it falls and take it away to a suitable place for burial, where they have a chance to keep it to themselves. Of this group, the *S. laevistriatus* are like deep-chested, thin-legged distance runners. When they come flying in at 30 kilometers per hour, they hit the ground running fast. As I have mentioned, they come at dusk, just barely before the crowd — or cloud — of thousands of others, most of which are tiny beetles that bury themselves directly in the dung. The *S. laevistriatus* have to be fast, not only because the competition for dung is intense, but because they need time to package it and haul it off. They are in danger while aboveground at a dung pile, because mongooses and birds such as hornbills and guinea fowl check out dung piles for insects to eat. I often saw mongoose tracks and the remains of larger beetles — the soft abdomen had been taken and the rest left. Dung piles are scenes of opportunity much like large animal carcasses, but they are laced with a high dose of risk.

· · ·

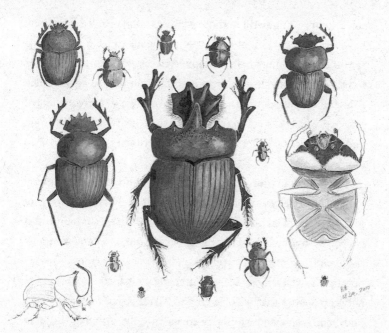

African dung scarab beetles I collected at elephant dung during research in Tsavo National Park in Kenya, along with two beetles from South Africa, Kheper nigroaneus *(top right) and an unidentified dung beetle species I collected in Kruger Park (center right). The large beetle at center is the elephant dung beetle* Heliocopris dilloni, *and to its left is* Scarabaeus laevistriatus. *The tiny one at lower left lives in the dung pile. Many of the others roll the dung away, and* H. dilloni *buries its ball directly beneath the pile. The larger beetles are black or brown; some of the smaller ones are metallic green or blue.*

IN AFRICA IT is hard not to be reminded of our beginnings. Almost at random, as I watched for dung and beetles on the ground, I found chipped stone, and one six-inch rock with many rough chips flaked off that was almost certainly an Acheulian hand ax (see my drawing on page 50) dating back perhaps a million and a half years. I picked it up, and I still hold it in awe today, not quite believing that it is what it appears to be, a hand ax probably used to cut into animal prey or carrion. The scene on the wall of a nearby

139

rock shelter reminded me that hominids, who evolved here, may have faced a situation at carcasses analogous to what the beetles face at elephant dung balls. It depicted a series of thin stick figures in full running stride — in form and function they were "distance runners" chasing down an antelope. Would these early men have grabbed what they could from the carcass as quickly as possible and then run? Or was their speed more critical in catching a live animal?

Fresh elephant dung balls are relatively solid, and when a *Scarabaeus laevistriatus* lands, its first priority is to run all over the pile in search of an already made ball. Stealing saves the time and energy necessary in making his own ball. But as in any competition involving physical strength and agility, body (muscle) temperature as well as body size are often deciding factors. Most fights between beetles that we observed in Africa lasted only a few seconds, and losers and winners were easily differentiated; the loser was the one who was flipped off the ball, and the winner the one who rolled it away. Immediately after each fight, we weighed the contestants and took their body temperatures with an electronic thermometer. To our surprise, the winners were not necessarily the largest beetles; they were those with the highest muscle temperature, several degrees centigrade higher than our own. These were the ones with the fastest leg speed, which in beetles is directly related to muscle temperature.

The ground under our feet was red clay, and I used water from our water bottle to make clay balls. The beetles ignored them until I dipped them into fresh elephant dung, after which they fought over them as readily as for the "real," beetle-made balls. This worked so well that Bart and I staged many fights without having to wait for fresh balls to be made, to the point that we were not always able to intercept the clay balls before they were rolled away.

Although the beetle with the higher body temperature is favored to win a contest over a dung ball, that heat comes at a price. The beetles can shiver while they make their ball, but if they shivered for the whole half-hour it takes, they could burn through their energy reserves. They are like long-distance runners who can sprint at the beginning of a race but much less readily at the end, when their energy reserves are depleted. Thus the beetles seek out others with balls as soon as they land, when they are still fresh, warmed up by their flight metabolism and thus better able to win contests. To avoid having their balls taken, the ball makers must try to stay hot, continuing to shiver to protect their investment. Not all of them could do so.

When there were no already made balls to be had, the *S. laevistriatus* rollers quickly got to work tearing bits of dung out of the pile with their pitchforklike front tibia and then patting the dung into a ball, their front feet all the while moving in a whir. They were able to move fast because they had arrived hot-bodied from their flight; the exercise created and stored heat. Some of the large beetles had body temperatures up to 113 degrees F, about 15 degrees higher than ours and that of most other mammals. Hot beetles worked fast and usually could finish a ball of about baseball size in five to ten minutes; then they would start to roll it away.

Their long slender legs moved fast as they ran and rolled the ball, and if they were still at about 108 degrees F, they could run at an average speed of 11.4 meters per minute on level ground. If their temperature was 90 degrees F, however, their running speed was only 4.8 meters per minute.

We ourselves were running out of time and resources in our fieldwork, and after we returned home I was left with a question. Under less competition, would the *S. laevistriatus* work at a more leisurely pace and not bother to maintain a high body temperature? We had found that one species that worked in the daytime, when there is little competition, did work more slowly and had lower body temperature. In part to answer my question, I went with my graduate students Brent Ybarrondo and James Marden to southern Africa. Unfortunately, we found no *S. laevistriatus* during the several weeks that we were in Botswana, the Union of South Africa, and Zimbabwe.

James instead studied a species of swift-running tenebrionid beetle in which the males run on all six legs behind females on the ground. Brent and I examined *Kheper nigroaeneus*, a dung-ball roller in Kruger National Park. Like most other dung-ball rollers, the parents in this species have only one offspring per brood ball, which the female tends underground for around twelve weeks. Like the nocturnal *S. laevistriatus*, these diurnal beetles fought at dung piles, and again the winners were the hotter ones. But if there was stiff competition at the fresh dung source, the beetles either left to search for another, less crowded dung pile or they made only a *small* ball, thus reducing their ball-making time and the chances of having their ball stolen. The drawback of this strategy, however, was that the little balls could not provide enough food to serve as brood balls; they were used only as food for the adults. (Because the female expects to rear an offspring only at a large ball, she probably chooses the ball rather than the ball maker.)

In my collection was one beetle that I found puzzling. It had a large, generally flattened body and superficially looked like *Pachylomera femoralis,* but unlike that species, which has been reported to roll balls, this one had none of the anatomical tools for ball making, namely, flanges on the front tibia and front of the head. Instead, it had hugely developed front femurs studded with sharp spikes on the front. It looked like a "degenerate" *P. femoralis,* and from body shape and structure I wondered if it might, as in bumblebees, be a species that had evolved by parasitizing a closely related host. From its shape I suspect that it may have evolved from being a dung roller to one that enters the tunnels of other species, then uses its hugely muscular front feet to pry the competition off their brood balls, the way *S. laevistriatus* pries competitors off freshly made balls.

IN THEIR GREAT diversity of specializations to harvest a similar resource, the dung beetles are an evolutionary laboratory. The first dung-eating beetles had the resource to themselves; all had an equal chance, and not much speed and skill were required. After competition made the resource more difficult to get and keep, specialists had an advantage. Getting there first was part of the winning combination to get dung from the dung-pile bonanza.

Most beetles are warm-weather flyers. As with the sexton beetles, which in Maine and Vermont are common only late in the summer and when it is warm, the dung beetles are highly seasonal and seem to be largely restricted to the tropics. I have seen only two dung scarabs in Maine, and both were on carrion. One might guess that they went extinct along with a main food source, the dung of bison, but cows took the place of those herbivores, and there are no dung scarabs at the droppings of cows, deer, or moose; in contrast, the dung of the equivalent bovids and antelopes in Africa is used up almost immediately by dung scarabs in the rainy season. In northern Europe, where the ancestral bovids have been exterminated and replaced by cows, there are no swarms of dung scarabs at dung pats — at least I saw not one scarab during two weeks in August 2011 when I worked as a cowherd in the Swiss Alps. Australia presents still another scenario. Here there is a tropical climate but no resident bovids until very recent times, when cows were introduced by Europeans.

The dung beetles' work has ecological significance. It fertilizes and aerates the soil and retards the spread of pathogens and disease organisms. But different types of dung beetles are adapted to handle specific kinds of dung in different seasons and habitats. Their specific roles in the ecosystem are difficult to determine, because we can't experiment by removing them. However, an almost continent-wide "experiment" from Australia answered many questions. The Australian entomologist and ecologist George Bornemissza, who was born in Hungary, collected beetles as a boy in that country. After obtaining his PhD at the University of Innsbruck, Austria, he emigrated and joined the zoology department at the University of Western Australia. One of the first major differences he noted between his native Europe and Australia was the large number of dung pats covering the ground where cattle were grazing, which he had not seen in Europe, where the dung decayed because of the moist climate in the north and because of beetles in the south. He realized that the native Australian beetles were not adapted to handle the dung of cattle. He therefore proposed to import beetles who were up to the job, and he started what would be a twenty-year work, the Australian Dung Beetle Project, for which he would receive the Medal of the Order of Australia in 2001.

Bornemissza, supported by the Commonwealth Scientific and Industrial Research Organization (CSIRO), searched in thirty-two countries for the most suitable dung beetle species to handle the cowpat problem in Australia. It was indeed a serious problem for two main reasons. First, the cowpats had a tendency to dry out and remain on the ground, in time reducing grazing land. Second, they were an ideal breeding ground for the very pesky bush fly, *Musca vetustissima*. If Bornemissza could find beetles that were able to stand the Australian climate and that would recycle the dung, it would solve two great problems in one stroke.

Introducing exotic species is always potentially dangerous, and all the beetles Bornemissza wanted to try out had to be raised in quarantine to make sure they were not carrying parasites that might become pests. Fifty-five dung beetle species were introduced into Australia, and they have become a proven success; they are now saving pastureland by increasing the health of the soil. The final report of the Queensland Dung Beetle Project concluded that the beetles are "undoubtedly worth many millions of dollars per year." The pesky bush flies have been reduced to such an extent that the "Australian salute," a flick of the hand used by generations of Australians to shoo away the fly, has become a "dwindling gesture."

Bornemissza has now moved to Tasmania, where he and Karyl Michaels have focused on the impacts of clear-cutting and burning of logging slash, which has led to a large decrease in the beetles that reproduce in decaying wood. One of those wood-boring species, now endangered, has been named after him, the stag beetle *Hoplogonus bornemisszai*.

THE DUNG BEETLES, despite their diversity and ability to handle almost any kind of dung, are like many undertaker beetles in that they handle only the fresh article. Once the dung dries out, as it usually does in the tropical dry season, the beetles stop work. They are active mostly during the wet season, when they can burrow into the soil. Their young develop underground in the dry season, and the signal for their reemergence is the seasonal rains, which soften the soil. All the dung deposited in the meantime would be left on the ground — were it not for termites.

Elephant dung is very coarse compared to any other dung I have ever seen, smelled, or felt. That is because elephants don't eat just young juicy grass shoots; they eat whole bushes. The result of digesting them can be stuff with the consistency of moist sawdust. You don't really notice how rough it is until a few thousand dung beetles have ravaged a pile throughout a night. By morning the beetles are done, and all that remains is a thin mat of fibrous frass about a yard wide. But when this woody stuff dries out, it becomes perfect fodder for termites.

Termites evolved from ancient cockroaches that ate decaying wood, incorporating bacteria and protozoa into their digestive tracts to help them digest the otherwise nonnutritious cellulose. Ensconced in a large log, the termites could eat slowly; there was plenty of food for the offspring, so the adults could stay home. Eating more wood merely enlarged their home, which they make by recycling their own feces. So the more inhabitants the merrier. Living a crowded but sheltered existence, termites had evolved from their cockroachlike ancestors by about 300 million years ago.

Even more than their cockroach relatives, termites shun the light for most of their lives. They fly from home only once, to seek a mate and start a family, which may become a colony of millions. But on average one new colony just replaces an old one, and each colony has only one reproducing female. So, since each colony sends out millions of males and females, successful reproduction (meaning raising offspring) is a lottery: if only one in a million females may win, 999,999 necessarily fail. Most of the colony residents stay all their lives in or close to their climate-controlled home. For foraging, they build long extensions of their home in the form of tunnels. Their fuel — cellulose from wood — is cheap and plentiful.

The termites' main potential pollutant is their feces, which contain the indigestible lignin from wood left over after they digest the cellulose. But it is this, their own feces, that they recycle as a building material for their homes and tunnels. I suspect, though, that termite feces contain some other ingredient(s), because the building material they make is remarkable, as I found out when I recently brought back a piece of a termite nest from the tropical jungle of Suriname. The material had the consistency of plastic and was to my surprise totally insoluble in water. This plasticlike material has not been researched very deeply, but I suspect that it would be found to lack the toxins, such as sex-hormone mimics, found in manufactured plastics. The termites' evolution-tested product might substitute for what we have invented, and it would use up lignin as well, which to us is a waste product when we extract the very useful cellulose from wood.

IV

WATERY DEATHS

As terrestrial animals, we automatically connect under-taking with burial, which implies a rooting to ground, usually in the place of living. But most of the globe is covered by ocean, where animals may die far from where they lived. Large carcasses like those of whales may sink miles into the cold, dark depths. Salmon may live most of their lives in the ocean, yet they come inland to die and be deposited in fresh water, and the major effects of their recycling are on land, not in the oceans where they lived. Watery deaths involve similar principles to deaths on land, but those principles are differently applied. They give us examples of how life adapts and glimpses of different worlds from those most familiar to us.

SALMON DEATH-INTO-LIFE

AT A SERIES OF SHALLOW WATERFALLS ON THE MCNEIL
River in Alaska, adjacent to Katmai National Park and Preserve,
the sockeye salmon migrating upstream to spawn in June, and the
chum salmon in July and August, encounter a gauntlet of grizzlies. These brown bears, *Ursus arctos horribilis*, are the largest in
the world, weighing up to 1,500 pounds. Thanks largely to Larry
Aumiller, who manages the McNeil Sanctuary, people who win a
select drawing to come here can see the bears within several feet.
They view them unprotected by any barriers, and they are not allowed to carry guns. No one has ever been attacked, because the
bears have become habituated to people and are not angered by
their presence. Besides, I suspect salmon probably taste better
than humans, at least as far as these grizzlies know.

The normally solitary bears gather at the McNeil River Falls
because the water is funneled to places where they can perch to
intercept the fish. The salmon have by then grown for two to three
years in the northern Pacific. From twenty to sixty-eight of the
huge bears may come to the falls at one time. Their great size is
due to their access to the rich salmon diet. During a good salmon
run, they become so glutted that they cease to eat the muscle meat,

instead pulling off the skin and eating only the fish's gonads, which are engorged with roe or milt. They may also eat the salmon's brain, another delicacy because of its high fat content. By the time they hibernate in the fall, they will have gained a couple of hundred pounds of fat.

One might suppose that the salmon not eaten is "wasted." From the ecosystem perspective, however, the bears' picky eating habits provide food for others. Wherever the bears are feasting, at the McNeil Falls or other places where salmon are caught, scavengers are there to take the leavings. In this case the scavengers feasting on salmon leftovers are swarms of gulls.

Various species of salmon run in Alaska's rivers and in many other rivers on the west coast of America. For most of them, the upriver fight is a one-way trip. Years earlier they came down the same river and then grew to adulthood in the ocean. They returned home to reproduce and then die. Indeed, soon after they entered the fresh water of their home stream, hormones kicked in to change their physiology and, in the case of the sockeye, also their appearance; returning sockeyes grow a large jaw and a humped back and turn bright red. After spawning, they experience a sudden, physiologically assisted aging; their tissues almost literally dissolve, and they die where they were born. The change in their appearance may be related to sexual selection, since many other fish also change appearance at breeding time, as do many birds. However, their seemingly premature death is more difficult to explain on the evolutionary principle of "survival of the fittest."

According to human standards and to our standard (simplified) ideas of so-called survival of the fittest, one might say there should *not* be an accelerated rush to death. However, according to most evolutionary logic, there is no point to living on after reproduction. Indeed, one might posit that at this point in human evolutionary history, our genes, on average, contribute to our demise, since we as a species continually collect more deleterious genetic mutations while having little or no natural selection that removes them. As we live longer and longer, medical costs will continue to rise. However, even by a strictly materialistic interpretation, we humans contribute much more than just our genes to future generations. This contribution is itself part of our genome as social beings in a complex world requiring skills to survive and prosper. The evidence for this interpretation is in our longevity. Our long postmenopausal life can be rationalized as being adaptive because old people, like old elephants, are able to pass on experience and knowledge to their offspring to help them survive and prosper. It seems to me that a similar argument can be made with regard to the salmon's seemingly premature death — that, as I'll explain, it is an indirect genomic mechanism of contribution to future generations.

After swimming perhaps hundreds, if not thousands, of miles upstream and spawning, a salmon has only a small chance of surviving the trip back to the sea, living another year, and making the trip back up to spawn. Therefore, rather than "save" anything for the very uncertain future, it might as well use everything it has now. And the only contribution that makes any conceivable difference is to ensure that the salmon's one reproductive effort is the maximum. This in itself would explain a lack of investment for future living, but does it explain what amounts to *suicide* rather than a gradual or unassisted decline?

Given that animals have had strong selective pressure to repair their wounds and avoid being eaten, we may wonder why these salmon literally give up and offer themselves to the predators/scavengers long before we surmise they must. Why should the salmon's migration to ocean and back be more difficult the second time as opposed to the first? Only one in hundreds may make it the first time, so why should the slim chances of making it a second time be less privileged? The relevant point is that the salmon presumably *could* feed after spawning and at least potentially recover for another try. But they don't. Instead, their behavior has two effects: it is in essence a starvation, because it *ensures* that they will exhaust themselves at some point. It also assures, however, that they will not eat their own or others' eggs and young. I believe that the second effect is the major selective pressure; self-imposed mortality helps the survival of their offspring. Recall that the fish are returning to their *home* stream and to the specific area in which they were born. This is the area to which their offspring will also eventually return and where their relatives are. And if not eating their own young and relatives is not enough selective pressure, another, even more indirect effect might enhance and would at least not detract from the selective benefit of offering themselves. Namely, this: the massive influx of their bodies helps create and maintain their ecosystem. The counterargument to the scenario I am proposing is that "cheater" fish who do not commit hara-kiri could potentially be selected for and replace those who offer themselves and benefit the group. In the case of the salmon, this didn't happen.

As I mentioned earlier, some of the salmon are taken by predators on their way up to the spawning grounds, but most are eaten on those grounds. The annual die-off of thousands, and in aggregate millions, of bodies creates a feeding bonanza greater than anything like that at the McNeil River Falls. Brown bears all the way up the West Coast and into Alaska feast on the dead and dying salmon, along with gulls, bald eagles, ravens, otters, crows, magpies, jays, and raccoons. These scavenging animals do what bears traditionally do in the woods, and the salmon are thus the "delivery packages" that bring nitrogen, phosphorus, and other nutrients from the ocean into the rivers and surrounding forests. Availability of nitrogen is a limiting factor in tree growth, so the salmon not only make big bears, they also help make big trees. In turn, the trees' roots, which hold the moisture of frequent and heavy rains, "make" the watershed and perhaps also the conditions needed for spawning.

The epic migrations of salmon and their spawning have always fascinated us, and the lives of peoples all over the world have depended on them. How much more fascinating they are, it seems to me, if seen as one of the most highly evolved death-into-life cycles in nature.

OTHER WORLDS

IN 1970 A SPERM WHALE CARCASS WASHED ONTO A BEACH near Florence, Oregon. Officials of the Oregon Highway Division, in consultation with the United States Navy, were concerned about the unbearable stench such a huge carcass would produce for a year or more and were at a loss about what to do at first. They decided to break up the carcass and thus facilitate its removal by scavengers. To accomplish this, they surrounded the whale with twenty cases (a half-ton) of dynamite. Soon after the fuse was lit, there was a stupendous rain of blubber chunks for 800 feet all around, one of which smashed a car a quarter of a mile away. These results, curiously, had not been fully anticipated.

Another stranded sperm whale (sixty tons), which beached near Tainan City in Taiwan in January 2004, also made headlines. This one was loaded onto a truck and taken to a university to be autopsied. But when the truck arrived, permission was denied. Later, on its way to a wildlife reservation for disposal of the carcass, the truck drove through the center of Tainan, where, on a busy street, the gases produced by internal decay caused the whale to explode. Aside from the nauseating gas, a shower of entrails and blood rained down on shops and people, and although the crowd tried to disperse, the hubbub stopped traffic for hours.

These mistakes were not repeated in 2007 at another whale carcass (seventy tons) that washed onto a beach in Ventura, California. This carcass attracted a huge crowd, but the Ventura County Parks Department had bulldozers dig a fifteen-foot-deep hole in the sand rather than fragmenting it by dynamite. The whale may, however, already have been punctured, as it (along with another whale around that time) was likely a fatality from a collision in the busy shipping lane of the Santa Barbara Channel. Unfortunately, most of the sand around the carcass washed away. Oil and rotting flesh leaked out, making nearby beaches uninhabitable.

We don't know how whale carcasses would have been disposed of in the Early Pleistocene and before. But beaching on land would have been rare, so scavenger specialists would probably not have evolved specifically to handle whale carcasses on land — just as we humans haven't worked out an appropriate protocol for this situation. Any stranded whales would have been used opportunistically by dire wolves, condors, and perhaps the American lion and saber-toothed cats who happened to be near.

THE NATURAL PROCESS of whale recycling presumably begins near the surface of the water. We know little about a whale's natural death, but we can imagine a scenario of what it may look like. Perhaps the whale weakens from old age and then drowns. I suspect that a weakened whale might easily become prey to orcas (killer whales), who hasten its death. After the orcas have taken their fill, the blood would attract large sharks, such as the great white, and various smaller sharks would come flocking to fresh meat. The whale's body cavity would be breached, organs removed, and the lungs deflated. What happens then?

The whale carcass begins to sink, drifting through a netherworld of dark, cold water populated by an assemblage of creatures that are specialized to live off the largess that comes down from above. These creatures seem bizarre to us because they are configured differently from those we know well. Some of the fish have light-generating organs, including one that resembles a lantern suspended from a stiff rod. Some have mouths that are larger than their bodies, with huge teeth. There are females who carry around tiny males that are like parasites embedded in their flesh, an adaptation that compensates for the difficulty of meeting a mate — something we take for granted in a world of light.

But these creatures don't catch all the manna that drifts down. Some parts of the whale continue to drift all the way to the bottom. Below a depth of 150 meters, photosynthesis cannot occur, so only animals, not plants, exist at lower depths. Those that have adapted to survive there either live on the largess from above or they catch and eat each other. Many are transparent. No light would be visible to us in this deep-water world, but the eyes of some of the animals are enlarged and especially well developed; those with some vision can more easily prey on those who see less and swim above them. Still farther down, where there is absolutely no light from above and no animal can see images, as we do by the light reflected from objects, the animals generate their own light. Prey animals obviously do not "want" to be seen, but they may need to be visible in order to be found by potential mates. At these depths beyond sunlight, there is a continuous light show of flashing and glowing blue lights that have different meanings, from (presumably) attracting mates to luring prey to faking out potential predators; one copepod has been observed to discharge its own light-generating matter (bacteria?) into the water to hide its location, much as some octopi conceal themselves by squirting ink. This is the world of the "engulfer eel," which hangs in the water and presents a long tail to make contact with drifting edible debris or swimming animals. It has a mouth big enough to swallow animals its own size. A forty-meter-long colonial jellyfish has plenty of surface area for contact with drifting food particles. Here lives the fangtooth, a grotesque fish with an appropriate name. It moves very slowly and uses sensory filaments extending from its body to detect nearby objects in the dark by touch or subtle movements of the water.

Finally, the whale, after sinking through strange dark worlds for many miles, comes to rest at the bottom. Here temperatures are near the freezing point, and bodies could potentially pile up forever in this refrigerator. But whales have been on earth in recognizable form since the Eocene, about 54 to 34 million years ago, and through all this time they must have been recycled, or the oceans would now be filled to the brim with their cold carcasses. Such a massive food bonanza as whale carcasses, drifting down to the ocean bottom over millions of years, would presumably have prompted a retinue of specialized scavengers to evolve to make use of them. Until recently we had no idea who these scavengers were or how they recycled the world's largest mammals.

Most of the oceans' ecosystems are ultimately dependent on the sun's energy captured at the surface. In the last decades, however, two new ecosystems have been discovered that suggest other possibilities for life. Deep in the ocean trenches, we now know, vents like belching chimneys spew out water heated to 400 degrees F and containing hydrogen sulfide (familiar to us as the rotten-egg smell), and some bacteria are able to use this chemical as an energy source. In this deep-water ecosystem, life is driven by chemosynthesis rather than photosynthesis. Shrimp and other organisms graze on the bacterial mass the way antelope graze on grass. Some of these bacteria have also evolved to live in symbiosis with animal cells. This is similar to the way chloroplasts evolved from algae living in symbiosis with cells and to the way bacteria evolved into mitochondria, allowing animals to live off plants or plant eaters.

In this recently discovered ecosystem of the "smokers," sulfide-eating bacteria feed worms, clams, crabs, and potentially many more organisms. A second newfound ocean-bottom ecosystem is fed by methane gas generated from "cold seeps." The methane is first captured by bacteria living in symbiosis with other organisms and feeding on the carbon compounds scavenged from them.

Beyond these two ecosystems is a unique third one, which depends on dead whales. Like the salmon traveling hundreds of miles upriver to die, those whales come from a different ecosystem, namely, the top layer of the ocean, which is driven by photosynthesis.

At depths below 2,000 meters there is almost no free oxygen, and temperatures range from 30 to 36 degrees F. In these conditions bacterial decay as we know it is either nonexistent or very slow. This point was proved in an unplanned experiment by the submersible vessel *Alvin*, built in 1964 and operated by the Woods Hole Oceanographic Institute, which takes two scientists at a time to great depths. In October 1968, while *Alvin* was being transported by a ship, a steel cable snapped, and the submersible sank 1,500 meters (5,000 feet). When *Alvin* was recovered ten months later, a cheese sandwich that had been left inside had not changed visibly, and someone was able to eat it. In those conditions, how would a 160-ton blue whale carcass be disposed of?

After being recovered, the *Alvin* was rebuilt, and since 1977 it has made hundreds of dives to advance our knowledge, especially of the deep hydrothermal vents at the midocean ridges. In November 1987 Craig Smith, a University of Hawaii oceanographer, was on a routine mission of the *Alvin*, using scanning sonar to explore the muddy bottom of the Pacific Santa Catalina Basin at a depth of 1,240 meters (4,070 feet), when he thought he saw a fossilized dinosaur. Instead it turned out to be the 21-meter-long skeleton of a blue whale. The crew was astonished to see it surrounded by mats of bacteria and clams. This sighting marked the beginning of the study of "whale falls," as they are now called. Since that first discovery, other whale falls have been found, and some have been deliberately created so that scientists can examine the progression of scavengers. The ongoing observations and studies show many specialist undertakers on whale carcasses, including previously unknown animal species.

We now know that, despite the low temperatures, the whale's flesh is fairly rapidly consumed by a succession of mobile scavengers after the carcass first settles on the sea floor. Numerous large, slow-moving sleeper sharks, which can live at great depths, move in, followed by swarms of eel-like hagfish, which burrow into the meat and absorb some of its nutrients directly through their skin. Rattail fish, stone crabs, and millions of amphipods (tiny crustaceans with laterally compressed bodies) also join the feast. This stage of feeding may be completed in months or a year, or up to two years with a very large whale. The bones take the longest to be recycled because their great bulk makes them resistant to access. Herman Melville, in *Moby Dick*, vividly describes the massive forty-odd vertebrae in a ninety-ton sperm whale as a "Gothic spire"; the largest vertebra "in width measured something less than three feet, and a depth of more than four."

After the soft tissues are consumed, a mat of bacteria colonizes the bones, and limpets and snails graze on them. The carcass becomes surrounded also by a dense mat of polychete worms, each around five centimeters long and superficially similar to centipedes. The worms cover the whole carcass in densities of up to 40,000 individuals per square meter. They take everything they can, and after they leave, an abundance of other species moves in. These thrive mainly on the nutrients remaining in the fat within the bones. Bacteria break down this fat in the absence of oxygen and produce sulfur dioxide as a byproduct; this, in turn, uses chemotrophic synthesis (as in the thermal vents) to produce organic molecules, much the way plants fix carbon dioxide through photosynthesis. As in a sunshine-driven ecosystem, animals in the whale-fall community live off the chemotrophic producers, some of which live inside the bodies of others, just as chloroplasts (from ancient algal symbionts) live in plants. Certain clams and tube worms don't need a gut because the chemotrophic bacteria they contain produce organic molecules directly within their bodies. Small *Osedax* "zombie worms" (*Osedax* is Latin for "bone devourer"), also with no digestive tract, tunnel into the bones to allow symbiotic bacteria to feed on fats, which the worms absorb into their bodies.

More than four hundred species of macrofauna (this category excludes bacteria) have been identified in whale falls, with at least a hundred at any one carcass. Tens of thousands of individual animals of many kinds may be at work decomposing a single skeleton at any one time. This stage can last ten years, and some think it may be close to a hundred years before the whale's decomposition is complete.

A whale fall is like a species-rich island. It is colonized by yet unknown means, in that the colonizers appear as if out of nowhere. Whale falls are habitats that contain specialists; they are hot spots of species diversity as well as sites of evolutionary novelty. The massive reduction in whale populations from excessive hunting in the nineteenth and twentieth centuries has surely spread these temporary "islands" of life far apart. We may wonder how far apart they can be before they are beyond the reach of colonizers, which would then die out.

THE FATE OF the gargantuan whale carcasses, with their specialized undertakers, contrasts sharply with that of the skeletal remains of minute marine plankton bodies, which also drift to the bottom. Like most organisms, whales are usually recycled down to the level of molecules, so they make their way back into new biological life. But in the case of some marine plankton, what endures after death is an astonishingly important geological building block, shaping the contours and geology and soil of continents, and hence what can grow on them, and also determining the earth's atmosphere and hence its temperature and the life it can support. In sheer volume, the aggregate mass of these marine plankton at any one time far exceeds that of all the whales. The most important present-day plankton already existed in the same form hundreds of millions of years before the first whalelike creatures plied the seas. The immortal remains of these plankton are chalk and the stone derived from it.

The story of chalk was originally and most famously enunciated in 1868 by Thomas Henry Huxley, the British naturalist probably best known as "Darwin's bulldog" because he vigorously defended Charles Darwin's thesis of natural selection. In a lecture titled "On a Piece of Chalk," delivered to "the working men of Norwich," Huxley noted that heating chalk causes the evaporation of carbonic acid and yields lime. So chalk is a carbonate of lime, the same substance as stalactites and stalagmites. Huxley examined thin slices of chalk under the microscope and found out that it was much more than a white chemical substance. Revealed were "hundreds of thousands of . . . bodies, compacted together."

These bodies, about one hundredth of an inch in diameter, have various forms; one of the most common looks something like "a badly grown raspberry" and consists of a number of nearly globular chambers of different sizes congregated together. Some chalk is made up of little besides these microfossils, a highly compacted mass of calcareous skeletons of one-celled, exclusively marine protozoa called *Globigerina*, one of the most dominant genera of chalk. There are about four hundred species of them in chalk that is 100 million years old or more, and thirty of these species are still living in our oceans.

In 1853, when a transatlantic telephone cable was being laid, the first samples of the ocean floor were retrieved, from a depth of around 10,000 feet. Mud dredged up from the bottom, when examined under the microscope, was found to consist almost entirely of the skeletons of a still-existing *Globigerina* species, along with the calcium carbonate skeletons of round, single-celled phytoplankton algae called Coccosphaerales, more commonly known as coccoliths. In short, the Atlantic mud, which stretches over a huge plain of thousands of square miles, is raw chalk.

Thomas Huxley is credited as being the first to see the single-celled plankton skeletons in his examination of ocean-bottom mud. One very important species, *Emiliania huxleyi* (Ehux for short), is named after him, and this species is by far the dominant coccolith in the oceans. It creates blooms covering tens to hundreds of thousands of square kilometers of ocean, giving the water a bright turquoise color that is visible from space. In the Upper Cretaceous, when hot magma from the spreading continental plates created large volcanic eruptions and greenhouse gases, global temperatures rose and melted the ice caps, and ocean levels rose to flood the continents some 600 meters higher than present levels. Ehux may have helped to reduce global warming at that time. The role of Ehux in contemporary global warming is currently being debated, since these algal blooms reflect light and heat and also remove carbon dioxide from the atmosphere to create the calcium carbonate plates that settle on the ocean bottom. Ehux has removed inestimable amounts of carbon dioxide from the atmosphere and locked it into chalk and limestone.

Chalk is found underground the world over; it underlies England, France, Germany, Russia, Egypt, and Syria, over an area about 3,000 miles in diameter. The chalk layers in some places are more than a thousand feet thick. Underground deposits of chalk are generally exposed at geological faults, but it is prominently exposed only on cliffs, perhaps most famously on the Cliffs of Dover, facing the English Channel.

In addition to its primary component, microfossils of sea plankton, chalk contains exquisitely preserved sea urchins, starfish, nautilids, and other mollusks, as well as plesiosaurs — in all, some thousand or more distinct species that plied the oceans in the Cretaceous. Sometimes the chalk contains streaks of black flint, which also comes from recycled animal remains, though its formation is not fully understood. The implication of Huxley's microscopic examination of a piece of chalk was tremendous: he was proposing that the vast areas of chalk-covered land had at some ancient time been at the bottom of an ocean. Huxley noted that about 5 percent of the Atlantic mud consisted of skeletons of silica rather than calcium carbonate. The silica is derived from diatoms — algae with a silica shell — and from the skeletons of sponges. From this one can deduce that the diatoms had to come from surface waters, where they could get light. In addition to making up what we call "diatomaceous earth," these organisms are thought to be a component of petroleum.

The origin of limestone is similar to that of chalk. It is a sedimentary rock consisting largely of calcium carbonate derived from the skeletal fragments of single-celled marine plankton, often mixed with the remains of clamshells, crinoids, and corals, which have created oceanic islands. The free-swimming larvae of corals attach themselves to a solid substrate and then build a calcium carbonate base that serves as a skeleton. When the coral animals die, this skeleton remains, and other individuals then attach themselves to it, building on the others and eventually forming a limestone reef. Limestone, and its metamorphosed form, marble, have been important building materials since antiquity. The pyramids are man-made mountains of limestone. The Romans were the first to make cement, by roasting limestone to about 440 degrees F to liberate a molecule of carbon dioxide from the calcium carbonate, leaving a powder that, when mixed with water, makes a binder for concrete.

The Roman Colosseum was built in part by sacking the city of Jerusalem, using the labor of an estimated 100,000 Jewish prisoners. But the construction would not have been possible without the then-recent invention of concrete. It was concrete mortar that set the travertine limestone, quarried twenty miles from Rome, and the marble for the seats and outer walls. Concrete was used to build the aqueducts bringing water for crops, life, and power to Rome. The material from which the Romans built their civilization almost 2,000 years ago we still use in public and private constructions of all sorts. I used concrete to bind the foundation rocks of my cabin in the Maine woods. We literally live on the remains of ocean life of past geological ages.

OUR HUMAN BODIES, too, are built from the lives of much earlier life. Our DNA not only holds the legacy of our own line reaching back to the dawn of life, it also incorporates the life of other lines of descent. The mitochondria in our cells — the powerhouses that allow us to burn carbon compounds and release the energy of the carbon-carbon bonds that we borrow ultimately from plants — are derived from ancient bacteria that set up life in our cells and that learned to live within their means by *not* multiplying, not using up more resources around them than their environment at the moment offered.

The ancient worlds live on even now. Many corals contain algal symbionts that give them their bright colors. The algae find a home in the coral and in return provide organic chemicals as food for them. Warm water temperatures may kill the algae, resulting in bleaching and eventual death by starvation. It was Charles Darwin who first hypothesized that coral reefs (and the islands that form from them) are created by the constant deposition of carbonate skeletons that do not disintegrate. Such reefs are one of the richest and most diverse — and most threatened — ecosystems on earth today.

The unrecycled remains of the sea world may have had the greatest geological and atmospheric effects, but on land the remains of the ancient world — mainly the unrecycled or incompletely recycled plant bodies that form peat, coal, and oil deposits — live on as well. In cold and oxygen-deprived conditions, plant remains are not recycled; they turn first into peat and then to lignite (which is less than 10,000 years old and still contains fibrous material), then bituminous or soft brown coal. With further aging, it becomes anthracite or "hard coal." (The origin of crude oil is still in question. One theory is that oil is *not* the result of incomplete decomposition of ancient plant life, mostly algae and zooplankton; the other, leading, theory is that it is.)

Coal came from the vast wetlands when the first amphibians crawled onto land and giant dragonflies flew through tropical forests of tree ferns and club mosses. Massive accumulations of mainly this plant life alternately sank and were flooded and then covered with sediments. Under high pressures and temperatures, these plant remains gradually turned to rock in a process that is still ongoing.

The vast amounts of coal that sparked and fueled the Industrial Revolution, made our massive population explosion possible, and that is still mined and burned now, was produced in the Devonian and Carboniferous periods, 360 to 290 million years ago. But much more ancient plant communities would by then already have remains left in anthracite. As this coal got folded deep into the earth by the colliding continental plates, to depths of 140 to

190 kilometers, it was subjected to a combination of intense pressures and temperatures that turned the coal into diamonds, the hardest naturally occurring substance.

Diamonds are to us a symbol of permanence and purity. Given their origin from life, they are also, it seems to me, an apt symbol of life's permanence, as well as its renewal through love. A diamond is a fossilized piece of life everlasting, forged in the context of the evolutionary history of the life of the planet. But if a diamond proclaims the preciousness of life, it is of all life through the ages, and not the bumper sticker version of potential lives of only one species now.

V

CHANGES

Culture is like the chalk and limestone made from the organisms of past ages under our feet. It's the residue of our knowledge, foibles, and aspirations that have accumulated over the ages. It's the nonmaterial life that we absorb into our brains through our eyes and ears, the way plants absorb nutrients through their roots and the stomates of their leaves and turn them into sugar and DNA. And, like limestone, this nonmaterial that we inherit and absorb has huge material implications for our own lives and for future lives. There is no clear boundary between physical and nonphysical recycling.

The mechanisms of cycling vary, but they all differ from the instant transitions of appearance only. A Nicrophorus *burying beetle changes "instantly" into what looks and sounds like a bumblebee, and some caterpillars make themselves look like a snake or a twig merely by changing their posture. Such fake transitions mask the fact that they haven't really changed at all. Real cyclings take a long time to accomplish and involve one body and mind metamorphosing into another with different physiology, behavior, and ecology. Physical transitions are routine in insects and amphibians, and to a lesser extent they also occur in other vertebrate animals.*

Before there was a science of biology, observations of natural transitions, such as that of a tadpole to a frog, undoubtedly led people to suppose that some equal magic might turn a princess into a toad or vice versa. And why not? The transitions in our own development from prepubescence to adulthood are based on processes roughly similar to those that cause a tadpole to metamorphose into a frog. Indeed, as developmental biology now teaches us, a human embryo is like a fish in the aquatic womb, where it makes the transition to what looks like a baby mouse before being born a human being. However, and this is an important point for us, we continue to meta-

morphose. We are cycled not only in the physical realm but also in mental and spiritual realms. And even more important, we are the only animal who has some control of our own metamorphosis, and of others', by our decisions.

METAMORPHOSIS INTO A
NEW LIFE AND LIVES

We make a living by what we get,
but we make a life by what we give.
— Winston Churchill

As our family steamed toward the New York skyline one morning in the spring of 1951 on our way to becoming Americans, my father was preparing me to see both the Statue of Liberty and *Kolibris* (hummingbirds). I was especially looking forward to the latter, and my joy on seeing the first one — a male ruby-throat — a few days later in Maine was tremendous. It took a while before I had one in my hand, though; although I was skilled with a slingshot, this bird was a challenge. But nothing prepared me for a catch I made a month or two later.

There are 339 recognized species of hummingbirds in the Americas. Many are major plant pollinators. They cause the miracle of transforming flowers from sheer beauty to seed-producing organs. I had no idea where in America hummingbirds lived, how many and what kind there might be, or what they might look like. But I had heard about a particularly marvelous hummingbird — the

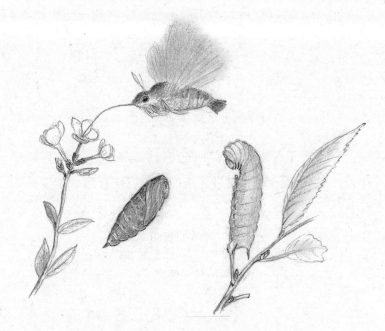

A hummingbird sphinx moth, Hemaris *sp., with its larva and pupa.*

tiniest one, the bumblebee hummer — and when I saw what I thought it should look like — tiny like a bumblebee but hovering rather than perching at flowers in our yard in Maine, its tiny wings whirring, its short tail flared as it darted from flower to flower — I wanted it. Badly. It was tame, so I ran into the house and got my father's insect-catching net, ran back, and in one swoop had it in the net. In this moment of triumph, watching it flutter behind the gauze, my excitement turned to surprise; I saw its pair of antennae — and knew then that it was a moth.

We identified it as a hummingbird moth, genus *Hemaris,* of which there are several species. This one had large eyes, a pale green back, and white fuzzy pile on its belly. Instead of a beak, it had a long proboscis, a tonguelike structure that when not in use is neatly curled up into a roll on its "chin." The moth belonged to a worldwide family, the Sphingidae or sphinx moths, sometimes called hawk moths. But aside from their behavior, which seems to mimic that of hummingbirds, and their physiology, which is analogous to a hummingbird's, one is struck by their beauty. The subtle color schemes of the sphingids are mesmerizing. Shades of gray mix with black and pure white, rich browns, yellows, purples, pink, ruby red, emerald green, in unimaginable combinations and designs due to their soft pelage of what appears to be hair but isn't. In birds, bright colors are used in sexual displays, but for these moths the color patterning serves as camouflage on backgrounds of bark or leaf, or as a way to startle a predator with a sudden display of false eyes. Except for the *Hemaris* hummingbird moths, most sphingids are nocturnal, and they all communicate by scent.

No sphinx moth at rest could be confused with a bird, but the moment it takes flight it looks like a hummingbird (though never a hawk!). These moths overlap with birds in size, and the largest have a wingspread of eight inches. They resemble hummingbirds in exterior form to fit into the same role for feeding on nectar by hovering at flowers. But their structural design could hardly be more different from a bird's. One has two legs with four toes on each, the other has six legs and no toes. One has a long bill and a long tongue, the other has a long proboscis, a sucking straw that can be rolled up or extended (in some it is more than twice the body length). One has a huge brain relative to its body size, the other hardly more than a little knot of neurons in its thorax and an even smaller one in its head. One powers two wings by an arrangement of muscles that pull directly on bones, the other has no bones but four wings. One has lungs that pump oxygen to its muscles by way of the blood, the other has no lungs and the blood does not transport oxygen. Hardly anything about them is the same, except that they look alike.

If we didn't know about sphinx moths, they would seem to be creatures from another world, yet they would be easily recognizable as a familiar form. But that is true in only one stage of their life. The moth has another, entirely different life, one that — if we were not already familiar with it — we would never guess was associated with the same animal. And if we examine the details of its metamorphosis, perhaps it isn't really all from the same genetic source. In this chapter I explore yet another way life transforms itself into another life, so that we can come to a more profound understanding of the concept and mechanisms of nature's undertakers.

LIKE OTHER INSECTS, a sphinx moth experiences a deathlike intermission of weeks to nearly a year, and potentially several years, between its two "lives." Had I wound the clock back about ten months, the hummingbird moth I captured that spring would have been a smooth-skinned, pea green animal with a huge gut, no proboscis, and no wings. It seldom moved except for its rapidly chomping, tiny, knifelike mandibles. It could crawl, but only slowly. As the moth's power and speed of flight are an adaptation, so is its larva's sedentary habit and slowness. The less it moves and the slower it crawls, the less likely it is to be seen by predators who hunt such prey relentlessly, cued by motion. Though of striking appearance when in your hand, in its natural habitat the larva is nearly invisible, blending in with the leaves it feeds on by subtle tricks of camouflage, including color matching, countershading, and a brown spot in its skin masquerading as leaf damage. The caterpillar remains firmly attached to a leafy twig, where it moves only a few inches each day to reach another leaf. It eats in such a way as to leave no leaf remnants, which would clue birds to its location; before it crawls to the next leaf, it chews off the remains of the first one at the petiole to obliterate its feeding tracks. If a predator, usually a bird, lands on the twig, the caterpillar rears back and assumes a stiff, lifeless pose like an Egyptian sphinx — hence the name "sphinx moth" caterpillar.

Of the many riddles of metamorphosis, one of the more speculative is why we have or need it. The default explanation is that metamorphosis is a necessary function of growth in reaching the adult stage, and that during this process the animal must develop through forms that trace the stages of its evolution. A tadpole is thus merely reliving its fish ancestry in order to reach the amphibian stage. Similarly, the gill slits and tail of the human fetal stage are also a retracing of that same early evolutionary path. These early developmental stages are useful guides to taxonomy because they are so conservative, whereas two totally different animals — the bird and the moth — show that convergence can lead to similarity.

On a practical level, consider a tuna and a whale: unlike a tuna, a whale fetus has fore- and hind limbs like those of other mammals, so we can deduce from its embryo that it is a mammal and not a fish, whose form it resembles. Similarly, a sphinx moth resembles a bird but is not a bird, and its larval stage, the caterpillar, resembles neither fish nor fowl nor amphibian. Larvae are, as Charles Darwin indicated, a good taxonomic tool. For a long time it was thought that barnacles, because of their hard calcium carbonate shells, were highly derived mollusks. However, Darwin showed that their larvae are free-swimming shrimplike creatures, so we now categorize barnacles as more closely affiliated with crabs and their allies than with snails.

The principle of developmental transformation that traces evolutionary change is called "ontogeny recapitulates phylogeny," an insight credited to the German biologist Ernst Haeckel.

It has often been harshly criticized, if not dismissed, because it does not apply universally. A phylogenic explanation might work well enough to distinguish insects from vertebrate animals, whose early embryos all look similar to each other but much different from other animals'. However, the early developmental forms of many marine invertebrates, such as sponges, starfish, and sea urchins, are wildly different from their adult forms. Octopi, on the other hand, which are derived from clams and their allies (earlier fossil forms had shells resembling those of snails), start life as plankton, but they have no metamorphosis with intermediary stages that look like snails. When octopi hatch out of their eggs, they look like octopi. Similarly, the principle of recapitulation of descent does not seem to apply very well to many other organisms, including insects, where the metamorphosis truly looks like transformation of one animal into an entirely different one.

Insects are an ancient group that arose in the Cambrian about 400 million years ago from aquatic crustacean-like ancestors. When they came onto land they brought along their protective armor, which also served as a skeleton and was modified into all sorts of forms. But in order to grow they needed to periodically shed this exoskeleton, which did not stretch unless it was soft. Every insect molts several times before it reaches the imago (adult) stage, and with each molt the body grows and may change shape slightly. That is, molting is the necessary step that permits transformation, but it does not necessarily result in metamorphosis. The most primitive insects make almost no change at all except in size; silverfish and springtails slip out of the egg as miniature adults. Similarly, very little change occurs from one molt to the next in grasshoppers (Orthoptera), true bugs (Hemiptera), and cockroaches and termites (Blattodea). Thus the question: why is there a radical, almost "catastrophic" body change in some orders of insects, such as the Lepidoptera (moths and butterflies),

Diptera (true flies), and Coleoptera (beetles), all of which have grub- or wormlike larvae? As I will explain, the radical change that occurs during metamorphosis in these groups does indeed arguably involve death followed by reincarnation.

Whenever two sets of very different genetic instructions operate, as in the metamorphosis of some insects and some other animals, the resulting incarnation is like a new species; as the larva's genetic instructions are turned off, those of the imago or adult turn on. But why are there two sets of instructions for two very different animals? The standard answer is that a caterpillar's specialized needs require different genetic instructions from those for the adult moth, which has other needs. But how could two genomes arise in one species? The view most widely accepted until recently was that this situation came about through a gradual process of natural selection, with different selective pressures at the two stages of the life cycle: strong suppression of the adult genes during the early stage and their activation at the correct time. However, a new theory claims that because the metamorphosis from maggot to fly or caterpillar to moth is so radical, with no continuity from one to the next, that the adult forms of these insects are actually new organisms. According to this proposal, sometime in their ancient heritage, when these animals were still aquatic and when all fertilizations were external, they hybridized with another species. They thus harbored a second set of genes, which could, given the right environmental conditions, be activated. In effect, the animal is a chimera, an amalgam of two, where first one lives and dies and then the other emerges.

At first glance, the idea of such metamorphosis originating from two different organisms living sequentially through recycling seems wildly improbable, and when the marine biologist Donald Williamson first proposed it, he was, as might be expected, ridiculed. But in fact the idea of chimeras incorporating genomes of other organisms is part of mainstream biology. When I was a graduate student in the 1960s working with the protozoan *Euglena gracilis,* it was known that while the nucleus of this protozoan contains its own genetic instructions, its body contains another, separate set of instructions in the mitochondria and yet a third in its chloroplasts. I raised these animalcules in the dark, feeding them sugar, acetic acid, and other organic compounds; they were scavengers. When I shone light on them, they turned into plants; because they had chloroplasts, they no longer needed to scavenge carbon compounds other than carbon dioxide from the air. Euglena and some other protozoa are able to transform themselves because along with their own DNA, the DNA in their mitochondria, derived from bacteria, allows them to break down and use sugar, and the third set of DNA, derived from algae, allows them to be a plant. Chloroplasts are archetypal blue-green algae that have adapted to live in a new environment — namely, inside other cells — and even to reproduce there. They have adapted, as all organisms must, by accommodation and restraint and response to appropriate stimuli in their environment. The chloroplasts' main adaptation was to *not* reproduce to the point of destroying their host.

In the above example only a part of another genome was incorporated to make the composite animalcule. But this process of DNA transfer happens all the time: when phage viruses infect bacteria, they often transfer genetic material from the infected cell into the genome of another, where it is incorporated and then multiplied ad infinitum as the cells divide and multiply. This process, called transduction, is a well-established experimental tool of molecular biologists. In the case of corals, mentioned earlier, whole cells from one organism live and multiply, but semi-independently, inside other cells. Similarly, the cells of some giant clams contain green algae, and like all green cells, these fix carbon dioxide into the carbon compounds that first build themselves and then also feed their clam host. And what applies to parts of cells living in other cells, and to whole cells inside cells of other organisms, applies also to whole *organisms* living in the bodies of other organisms, such as protozoa and bacteria living in the digestive tracts of termites, elephants, and numerous other members of the animal kingdom. Such symbioses also extend to the organization of ecosystems, and ultimately to the interdependence of millions of organisms — the whole biosphere of earth.

The idea that protozoans acquired useful genetic instructions from algae, and termites acquired other useful genetic instructions from protozoa, is a concept no wilder than that of humans incorporating domestic animals and plants into our society, giving us new genetic instructions for making McNuggets and French fries. What's different is the level at which the new genetic instructions operate.

Regardless of how it came about, there are indeed two very different sets of genetic instructions at work in the metamorphosis of some insects and some other animals, and these are as different as different species, or even much more so. They thus represent a reincarnation, not just from one individual into another, but the equivalent of reincarnation from one species *into* another. How the two coexist in the same organism without creating a garbled creature that is "neither fish nor fowl" is a potential problem regardless of how those two genetic instructions originated. The solution is that most of one body dies and the new life is resurrected in a new body. It happens roughly this way in all insects. When my hummingbird sphinx moth caterpillar had grown to its full size, it left the cherry tree where it had fed all its life to wander about on the ground and then burrow into the soil. There it created a crypt for itself; lying there motionless in the dark, it eventually shrank, shed its dead skin, and turned into a mummylike shape with a hard covering. As its organs dissolved, its insides turned to mush, and most of its cells died. However, some groups of cells, named "imaginal disks" (from "imago"), remained. These, like the buds on a plant that can grow into a twig and the twig into an entirely new plant, are like seeds or eggs generating new organs. During this apparent "resting" or pupal stage, the disks secreted enzymes that destroyed the larval cells and incorporated the proteins and other nutrients from those cells into themselves. Eventually all of the larval cells were replaced, and the new cells assembled in an orderly way to produce the moth. As with most of life, this process followed specific instructions encoded on genes that directly affect physiology.

In us the process of transformation is the same, but something new is added. First, the process of change is gradual and extends throughout our lives. Second, it is not just the genes talking; it is also the brain that, in thoughts and ideas, can almost literally cause reincarnations, in others as well as in ourselves.

BELIEFS, BURIALS, AND
LIFE EVERLASTING

*I have no doubt that in reality the future will be vastly
more surprising than anything I can imagine. Now my
own suspicion is that the universe is not only queerer
than we suppose but queerer than we can suppose.*

—J.B.S. Haldane, *Possible Worlds*

*In our family, there was no clear line
between religion and fly fishing.*

—Norman McLean, *A River Runs Through It*

WE MAY THINK OUR SPECIES GENETICALLY UNIQUE, AND IN-
deed it is, as every species is. But the mix of our DNAs is really an
amalgam of all life's DNA, and in many and varied ways that mix
reaches back to a common origin in the dawn of life. One example
of most recent common origin comes from our hunter ancestors,
whose skill and knowledge were pivotal, as we've seen, in the recycling of animal carcasses. Since those carcasses were derived from
formidable live animals, which the hunters had to get to know
well in order to hunt them effectively, we became empathetic. We
learned that the precious, mysterious gift that we call "life" may

disappear suddenly when the animal is punctured with a spear or arrow. In no area did we know less and need to believe more than in that period after death, when a body is little changed and yet suddenly bereft of life. Where has "it" gone, and where did it come from and why? We invented stories about human creation to try to make sense of our life and our fate, stories that specified our relationships to each other and to the earth, which then nurtured our morality. The knowledge to create these stories was short then, but the belief anchoring that knowledge had to be long. Metaphors helped explain the unknown in terms of the known. For the metaphors to seem true, they had to touch truths of our existence, and if they made us feel good they were more readily accepted.

To the Egyptians, the dung scarab beetle (probably *Scarabaeus sacer*) represented Khepri, the sacred scarab that rolled Ra, the sun god, up into the sky in the morning. Ra, believed to be the creator of all life, created himself out of nothing every day and was rolled across the sky, then returned back to nothing in the underworld at night. Scarab models were made by the millions as amulets and were placed on the heart of a mummified corpse in its preparation to enter the afterlife. Further instructions for human afterlife appeared in what came to be called the "Books of the Dead" (which the ancient Egyptians called "Books of Coming Forth by Day"), vignettes in hieroglyphics on papyrus scrolls illustrated with pictures of people, animals, demons, and gods. These papyrus scrolls accompanied the mummified corpse with its scarab beetle on the heart and were intended to instruct the spirit for continuing the earthly pleasures.

The most famous vignette, preserved in exquisite detail, is of a man named Ani, who lived at the time of Rameses II, around 1275 BC. We see Ani and his wife bowing toward the gods as his heart, the presumed seat of intelligence and the soul, is weighed by the jackal-headed god Anubis. Ani's soul is instructed to speak to his

heart. The feather of truth is a counterweight on the other side of
the scale. Toth, the ibis-headed god of wisdom, records the verdict.
Ammit, "the Devourer" (a monstrous chimera that is part croco-
dile, lion, and hippopotamus), awaits the outcome of the weighing, which will determine whether Ha, Ani's soul, will continue
to experience earthly pleasures during its daily journeys out to
Ra, the sun god. After making his daily rounds, Ha returns to the
mummified body at night. If the weighing of Ani's heart tips the
verdict to indictment, Ammit will swallow his soul. The Egyptians
believed they could influence the gods, and they had to adhere
to rules, practices, and conventions to prepare for their afterlife.
Those beliefs were strong enough to build the pyramids, whose
purpose was to facilitate the afterlife of the powerful people who
could afford the costs of construction. But the pyramids were also,
as the ancient Greek historian Herodotus notes, emblematic of a
time of horror for the masses, who were enslaved to build them to
ensure others' afterlives.

We don't believe this story of the afterlife anymore, in part because we understand it to be a toxic one that robs the poor to feed
the rich. We want everyone to have the same chance at life and
happiness. The problem, however, is that with so many religions,
by definition every believer of just one is a heretic to many others.
Most religions recognize this serious problem, and the traditional
remedy is conversion to the "one" religion, if possible, and imposi-
tion of that religion, if not.

For the ancient Egyptians, as for other cultures, ideas about
immortality were related to religion. These ideas often included a
belief in recycling in the universe as it was configured at the time
and that often involved, as is still the case in some areas, large car-
rion-eating birds. In the Tsawataineuk tribe in Kingdom Village,
British Columbia, a chief's soul returns to the village in the form of
a raven. The raven is still a powerful symbol of the afterlife, as the

letter from my friend at the beginning of this book attests. After I received that letter, another friend of mine told me he was trying to figure out how to get eaten by ravens after death: "I'm going to get cremated and have my ashes mixed with hamburger and fed to the birds." In ancient Egyptian beliefs, the mother goddess, Mut, was a griffon vulture, the medium to birth into another world. However, the dung-ball-rolling scarab beetle played an even more important role in beliefs about the afterlife. The dung beetles' life cycle apparently served as nature's verification of the afterlife and provided a model for humans of ways to prepare for it.

As I mentioned, the beetles bury themselves in the earth to rear their offspring. People planting or plowing land may have found their apparently lifeless pupae with the rigid impressions of the legs and other body parts pressed to the sides. They would have seen no internal organs, only the apparently lifeless body encased in a shell that included food for the animal's future life after metamorphosis. Observers then as now would have seen how one day a live beetle — a shiny new incarnation — emerges from this apparently lifeless pupa, comes up out of the earth, and flies away. They would have noted that this new beetle really "is" identical (in appearance) to the one that burrowed down into the soil a year earlier. The ancient Egyptians thought that this beetle had only one sex, which must have been an offshoot of the belief that a live one was resurrected directly from a dead one.

A civilization that had the means and power to build temples and pyramids, make fine fabrics, and fill libraries, one in which animals were configured as gods and beliefs were powerful enough to induce the building of pyramids to secure the afterlife, would have examined dung beetles and known some of their habits and life histories. They wanted to know about these animals that were relevant to achieving the afterlife.

The ancient Egyptians managed to weave an amazing number

of facts from nature into their creation story, but they had it all wrong: the proverbial devil is in the details. We now possess new knowledge of dung beetles and of much more, and we are writing a new creation story. To achieve the afterlife we no longer need to wrap the human body to make it look like a scarab beetle pupa, nor provide it with food in a dark, concealed chamber with a long tunnel (such as that dug by the scarab beetles) leading in or out so that the eventually resurrected life could fly and frolic.

The ancient Egyptians' beliefs concerning the recycling of human remains to achieve an afterlife are striking, but they are no more imaginative than those of other, earlier people, who were similarly ignorant of what went on beyond the flashy and arresting façade of nature. The first civilization as we know it in terms of cities, monumental structures, and centralized activities arose more than 2,000 years ago in what is now Iran. Before settling in cities, the people in that area were hunters living in villages. They probably worshiped vultures, ravens, eagles, and cranes or at least were impressed by these large birds. Vultures and eagles would have used the animal remains that were regularly available on village refuse heaps. These birds were apparently emblematic; we know their wings were used in ritual dances that might have been celebrations of life and death. Wall decorations at Çatal Hüyük, a Neolithic town in Anatolia, from 4,000 to 5,000 years earlier, depict almost life-sized vultures with short necks and neck ruffs (probably cinereous vultures, *Aegypius monachus*) feeding on headless human bodies. The anthropologist James Mellaart, who excavated the site, considered this depiction "proof of burial." Another wall decoration shows two griffon vultures (*Gyps fulvus*) with human body parts. The dwellings contained human skulls and sometimes jumbles of bones from incomplete skeletons. Did the people have places where bodies were deliberately set out for the vultures? If so, then the birds would have left de-fleshed skulls and some bones, perhaps those interred in the dwellings. Skulls found in Jericho had cowry shells inserted with clay into the eye sockets. Perhaps they were kept as mementos of the departed.

One of the Çatal Hüyük murals shows a human swinging something around his head. Mellaart thought that the person was trying to chase the vultures away. However, two vulture experts, Ernst Schüz and Claus König, posit that the person is trying to *attract* vultures. They base their hypothesis on observed customs in Tibet. One of the first Europeans to enter Tibet, the German explorer Ernst Schäfer, reported in 1938 that vultures there had been conditioned to *approach* when the *ragyapas* — professional body dissectors — swung a sling; the *ragyapas* would then distribute the body parts for quick removal by the vultures. When the birds had finished feeding, the *ragyapas* returned to crush the bone remains until almost nothing was left. This sky burial was a convenient, fast, and inexpensive way to dispose of the dead, and ideas of the afterlife could naturally then be incorporated into rituals and religious customs.

Vultures, ravens, and eagles soaring high in the sky would eventually be seen as mere specks, which would then disappear from sight. When these birds descended from the heavens in great spirals, with the wind fluttering through their great pinions, and took the bodies of the departed, it could have seemed logical that they had come from and would return to the home of the spirit world, carrying something vital.

MOST OF US want to remain part of the physical world for as long as we can, and we want another life we can believe in. The strength of our belief in another life depends on what we think we know. Few of us question the nature of the familiar world around us. And yet modern science is revealing our physical world to be more and more incomprehensible and mysterious the more we try to understand it. Most of us are consciously aware of our direct connections to the biological world and how they link us to history and time. Yet as the physicist Stephen W. Hawking explains in *A Brief History of Time,* ever since Albert Einstein challenged the notion of absolute time in 1905, we have had only a vague notion of what space is. We don't even really know what time is, yet it affects all of space and hence all matter. From a physicist's perspective, the universe is "curved" and has no beginning and no end. As a result, asking what came before the Big Bang could be meaningless because, as Hawking remarks, "It's like asking what lies north of the North Pole."

The little that we know brings some of our perceived connections to the physical world into the realm of metaphysics, and current science affirms the notion of mysterious connections. A note by Adrian Cho in the May 4, 2011, issue of *Science* reports that a $760 million NASA spacecraft mission has confirmed Einstein's theory of general relativity, "which states that gravity arises when mass bends space-time." Get it? I think I do: namely, the universe as we know it is a function of time, but we do not understand time, mass, space, or gravity. But that is what we are made of, what we are a part of. Nature is indeed incomprehensible at that deep level: there is more in our connections to it than meets the eye — and more than may ever be configured by our brain, even with its hundred billion neurons. I try not to be a sucker to our natural tendency to seek pleasure and satisfaction, which causes us to believe almost anything that makes us feel better and then deem it "right." But I cannot exclude the possibility that there may be other dimensions to the world aside from the familiar ones and that something lives on beyond my physical self. If so, when I pass on, it will be a celebration for some other beginning and not an end. Even if that is not the case, I have lost nothing and gained much.

Just as space-time connects the cosmos, and the molecules that make up our bodies connect us to past exploding stars, we are connected to the cosmos in the same way we are connected to earth's biosphere and to each other. Physically we are like the spokes of a wheel to a bicycle, or a carburetor to a car. The metaphor that we are part of the earth ecosystem is not a belief; it is a reality. We are tiny specks in a fabulous system, parts of something grand. We are part of what life has "learned" from its inception on earth and has genetically encoded in DNA that will be passed on until the sun goes out.

Beyond the most obvious physical-biological connections, we are an amalgam of past lives. This is true for all animals, but it seems especially relevant to us because we can in part consciously direct the trajectory of this inheritance. We know from personal experience, as well as from cognitive science, that we are what we experience and remember; we are a symphony of experiences. Almost every significant turn or change of direction in my life had a mentor behind it — someone who cared and to whom I was bonded and who opened my eyes or instilled spirit.

During my first year as a runner, when I was a junior at the Good Will School in Maine, I was mediocre at best. But by my senior year I had made a dramatic turnaround. The first meet that year was against the much larger Waterville team, and this time we faced their varsity, not the JV team we had raced against before. I won the race, and we trounced them. I was also first overall in our second race, against Vinalhaven, and we again trounced the competition. In each of the next seven meets I was the first man in. How was this possible? What had happened in the intervening year? I think I know: I was no longer the former Bernd Heinrich. Even my body was not the same; it now held the life spirit of a man named "Lefty" Gould.

Lefty was the postmaster of the one-room post office in the

town of Hinckley. I saw him twice a day when I brought him the school mail in a leather pouch. After he had removed the contents and inserted the incoming mail, I carried the pouch back to school and deposited it at the administration building. To Lefty I was not a bad kid, even though I was a mediocre athlete, had criticized my housemother, splashed red paint on the water tower, earned bad grades, and been booted out once. He was on my side, and he saw that I liked to run, just to be running. He, on the other hand, could barely walk. Whenever I came to the post office, he leaned on the sill of the window through which we exchanged the mail and talked to me as though I was someone who had worth. I think he saw me as an underdog who had gotten a raw deal, as he had, although he would never suggest such a thing. Lefty told me that he had been on his way to becoming the welterweight boxing champion of the world, and I had no doubt that he was telling me the truth. He told me how many pushups he used to do per minute, how many miles he ran every day. But fate intervened; he fought with the army's Eighty-second Airborne Division in Europe and North Africa and had one leg almost blown off in combat. It was a miracle that enough of it was saved (by a German doctor after he was taken prisoner) that he could, just barely, walk despite all the metal in his body. Sweat would roll off his forehead as he told me of his experiences in the war. I could not believe he was telling it all to me! I started running harder, faster, longer, even if it hurt, to show Lefty what I could do. He would never know, or even suspect, that part of his spirit would live on beyond his death. But it does. His belief in me and his mentoring are an inheritance from him. Every good race I have run, every running record that I have set, traces back to my last year in high school. Through our bonding, Lefty unknowingly set my wings and pointed me in a direction that led me to college and then to the opening up of the world.

We leave a legacy through our relations with people, mostly our parents and those who become in some way close to us. We are given much but must also receive actively. My father wanted me to carry on his lifelong collection of ichneumon wasps. At the time that seemed like becoming an extension of him, not a real interest in me. Yet much of him is in me. He gave me a masculine, vigorous love of nature, which at this moment is being expressed in my writing this and was a factor in all my previous work. It is the end result of countless excursions into the woods and fields collecting his wasps, listening to his stories, chasing rare birds in far-off, exotic lands, his taking me for a year into the African bush and its jungles. I disappointed him in not becoming an ichneumon wasp taxonomist, but deliberately or not, I took what he did offer.

The more I thought about it, the more I realized the obvious. We are not just the product of our genes. We are also the product of ideas. The shape of my body, the very oxygen-carrying capacity of my mitochondria, the physical circuits in my brain, and the chemicals that move me were in part shaped, if not determined, by others' ideas and thoughts. Ideas have long-lasting effects on us, as surely as, if not more than earthquakes, droughts, rain, sunshine, and other quirks of nature.

In springtime I walk on the snow crust formed at night after the daytime sun has thinned it to a fragile wafer; the ravens fly into the tall pines and build their nests of freshly broken-off poplar twigs and line them with deer fur and lay blue-green eggs. After the snow melts away, a riot of flowers — purple and white trilliums, sky blue hepaticas, yellow and blue and white violets, and snow white star flowers — bloom suddenly and disappear just as quickly. Meanwhile, ovenbirds call at dawn, the hermit thrush pipes at dusk, then the woodcock sky-dances over the clearing, and the barred owl hoots its maniacal cries from the deep woods. Summer brings the tiger swallowtails sailing through the woods and the fuzzy bumblebees to the yellow goldenrod in the fields. Come fall the Red Gods call me to the hunt of the rutting white-tails, and I look forward to the tranquility of drifting snowflakes covering all in white and sealing it in for another year, leaving a palette for the tracks of the tiny shrew and the mighty moose. Tiny kinglets with crimson and bright yellow crowns cavort with nuthatches, and brown creepers and chickadees flit among the red spruces, where they shelter from howling winds in blinding blizzards as the winds whip the trees. It's all in there — the Life — and I experience it and remember it and so become a part of it. You can't argue with nature. It is the primary context for living and for everything alive.

IT WAS NOT easy to give a suitable answer to my friend's letter. I could not receive a human body from a distant state and then lug it into the winter woods and leave it naked for the ravens; they might not show up for weeks. There are laws about proper burial, and such behavior would be illegal, so I could decline his request with a clear conscience. Still, my friend had a point. What better opportunity than death, not to sanctify an end but to celebrate a new beginning? What better time to confirm in ritual a model of the world as we know it, see it, feel it? I had no solution to offer him, and the problem of what to do nagged at me.

Here is a typical modern commercial burial: it starts with the naked body lying on a steel table, where the embalmer drains the blood and injects the body with a very toxic chemical — formaldehyde — that prevents decay. It is then placed in a metal casket and sealed so that no formaldehyde can leak out, as though it were hazardous waste at a landfill. Then "it" is added to millions of others, eating up more space every year — space that is kept largely free of flowering plants but instead is a monoculture of cropped grass, sometimes with cut flowers brought in that have been grown in a greenhouse. In the United States alone, the burials in our 22,500 active cemeteries annually eat up 30 million board feet of hardwood lumber, more than 100,000 tons of steel, 1,600 tons of reinforced concrete, and nearly 1 million gallons of embalming fluid.

Cremation was once an excellent sendoff. We can imagine it as a dramatic ceremony conducted at night at the edge of or in the forest, where plenty of wood was readily available. The ashes of the deceased were collected in an urn and then buried. Modern cremation, though, is not a ceremony, nor does it respect our home, the biosphere. It is more like a disposal. Vaporizing the body by fire creates emissions of toxic chemicals too numerous to list; modern industrial crematoria account for 0.2 percent of the global emissions of dioxins and furans, making it the second-largest source of airborne mercury in Europe. The amount of fossil fuel required to cremate the North American crop of bodies each year has been estimated to equal what an automobile would use in more than eighty round trips to the moon. Cremation is therefore a hugely expensive means of disposal. "Natural" or "green" burials, which are more personal, natural, and inexpensive, are increasingly recognized and practiced. (Anyone interested in green burial can find the most current information on the Internet.)

We deny that we are animals and part of the wheel of life, part of the food chain. We deny that we are part of the feast and seek to remove ourselves from it, even though we kill and consume animals by the billions and permanently remove the life resources for many more. But not one animal is allowed to consume us, even after we are dead. Not even the worms. We need a new creation story that connects us to nature and to others, one that can give us strength — that can make us real rather than rich. Nature, religions, and science coincide on the real: kinship with each other and with the mountains and prairies, oceans and forests. I am talking about beliefs built on facts that we all can agree to and that transcend individual deaths.

• • •

How would I like to be buried? I can barely plan for the next hour, so planning for decades ahead is a stretch. Sometimes I know more of what I don't want than what I want. I'd refuse formaldehyde because it is a biocide — it kills. It would hurt me as much then as it would now. But they could take any part of me and let it live on in another man, women, child, or Labrador retriever who might need it. If there was no human recipient, they could give my heart to the ravens, who have given me much. Some beer, a banjo, and maybe a guitar or two at a small ceremony in the woods would be nice. I'd want maybe a half-emptied bottle of Scotch whiskey by my side for the sendoff, the singing of "The Maine Stein Song," and a spokesperson to deliver my nod of acknowledgment to the great gift I received from Lefty. My tattered running shoes from high school, which I saved because they carried me where I never thought it possible to go, would do just fine to send me on my way, probably in a pine box under a tree.

WRITING THIS BOOK was a strong inducement for me to think about my origins and my fate. My highest aspirations, when I thought about belonging to something greater than myself, used to be an ecosystem. But because of the electrifying consciousness expansion that we are experiencing through modern technologies, I think we are within reach of seeing and feeling the whole earth biosphere. The world, not merely our neighborhood, is now our common reality. Nature is the ultimate standard of reality, and from what has been revealed so far, I see the whole world as an organism with no truly separate parts. I want to be connected to the grandest, biggest, most real, and most beautiful thing in the universe as we know it: the life of earth's nature. I want to join in the party of the greatest show on earth, life everlasting.

POSTSCRIPT

Writing this book morphed into a wide-ranging exploration of biology, conservation, human origins, and ethics. Publishing *Life Everlasting* revealed errors and holes in my knowledge but also led to discoveries. Readers pointed out the first, and I stumbled upon the second. As to the first, I fess up: In the book, I called blowflies (Family *Calliphoridae*) "botflies" even though I had called *Calliphorid* flies either blowflies or blue- or greenbottle flies since I can remember. But for the text of a book for general readers, "blowflies" sounded too colloquial and perhaps off-putting, and "Calliphorids" seemed too technical. Nevertheless, green- or bluebottle flies, named because they look like glass of these colors, are not "botflies." (This term is reserved for *Oestridid* flies, whose larvae are internal parasites.)

Perhaps ironically, despite my ignorance of *Calliphorid* flies (over a thousand species worldwide), I believe I made a scientific contribution to our knowledge of the group in the aftermath of writing *Life Everlasting*. It came to me because of my lack of expectations and from naively stumbling around motivated by curiosity. To my utter surprise, shortly after the book was published, while I was observing *Calliphorid* fly larvae (maggots) devour a raccoon carcass, I saw tens of thousands of the tiny white wiggling animals more or less simultaneously leave the carcass in one slug-like mass. Most curiously, not only did they leave at nearly the same time and in one mass, they also went in the same direction. It seemed miraculous to me that so many maggots could have an apparent, unanimous goal, and I could see no reason for their be-havior. I was hooked.

As almost always with the seemingly miraculous and mys-terious, the phenomenon can become almost mundane after you have the answer and understand it. I shall here defer risking the possibility of instantly rendering the maggot mystery mundane to you, but I hope it will still seem at least a little miraculous to the reviewers of my study of the maggot behavior of the blowfly/ greenbottle species (*Phormia regina*) in which I observed it. I have submitted my study for publication in a scientific journal (*The Northeastern Naturalist*), and peer reviewers will determine its worth.

As a result of a similar mix of naïveté and partial knowledge, while watching burying beetles earlier, I had (as already described in the book) stumbled on another undescribed mechanism. In this case, it was how one species' (*Nicrophorus tomentosus*) bright or-ange and black back morphed to *yellow* in a mere instant to mimic a bumblebee's. This observation, extended to become a study, has since been published in the above-mentioned journal.

ACKNOWLEDGMENTS

Writing a book is for me an adventure into the unknown. It starts from a point of familiarity, a background accumulated from the experiences, works, and influences of countless past and present lives. I could never hope to formally acknowledge it all. There is always the nagging worry of leaving out or not properly acknowledging those who have made a difference, especially in providing new insights and information. The best I can hope to do is remember some people with whom I have had recent conversations. Among those I thank Stephen T. Trumbo, Derek S. Sikes, John C. Abbott, and Alfred Newton, who kindly helped with my many questions on burying beetles. Barbara Thorne, Rudolf Scheffrahn, and Alison Brody answered questions about termites. Beth Rosenberg and Tom Griffin suggested the inclusion of salmon and generously invited me to a ringside seat in Alaska to observe them. I thank Baz Edmeades for channeling my views about ancient scavengers into a direction new to me. Rachel Smolker did the same on the recycling of trees in the context of industrial-scale logging. Richard Estes offered views of African wildlife. William Jordan and Janice Cahill pointed me to quotations and made useful suggestions. I am grateful to Sandra Dijkstra and Elise Capron for their continual interest and encouragement throughout this project, and to Peg Anderson, whose meticulous and insightful attention to details smoothed the way. Last in line but first in importance I extend my sincere gratitude to Deanne Urmy, who saw it first and saw it through with wise counsel to the end.

FURTHER READING

Many of the headings I have chosen are arbitrary, especially those that do not subsume primary research articles. I make no attempt to offer a survey of the relevant literature, which would include thousands of references. Instead I hope to provide an introduction to the topics through a modest list of sources I've found helpful and interesting.

Beetles That Bury Mice

General Biology of Burying Beetles

Fetherston, I. A., M. P. Scott, and J.F.A. Traniello. Parental care in burying beetles: the organization of male and female brood-care behavior. *Ethology* 85 (1990): 177–190.

Majka, C. G. The Silphidae (Coleoptera) of the Maritime Provinces of Canada. *Journal of the Acadian Entomological Society* 7 (2011): 83–101.

Milne, L. J., and M. J. Milne. Notes on the behavior of burying beetles (*Nicrophorus* spp.). *Journal of the New York Entomological Society* 52 (1944): 311–327.

———. The social behavior of burying beetles. *Scientific American* 235 (1976): 84–89.

Scott, M. P. Competition with flies promotes communal breeding in the burying beetle, *Nicrophorus tomentosus*. *Behavioral Ecology and Sociobiology* 34, no.5 (1994): 367–373.

———. Reproductive dominance and differential avicide in the communally breeding burying beetle, *Nicrophorus tomentosus*. *Behavioral Ecology and Sociobiology* 40, no. 5 (1997): 313–320.

———. The ecology and behavior of burying beetles. *Annual Review of Entomology* 43 (1998): 595–618.

Sikes, D. S., S. T. Trumbo, and S. B. Peck. Silphidae: large carrion and burying beetles. Tree of Life Web Project, http://tolweb.org (2005).

Trumbo, S. T. Regulation of brood size in a burying beetle, *Nicrophorus tomentosus* (Silphidae). *Journal of Insect Behavior* 3 (1990): 491–500.

———. Reproductive benefits and duration of parental care in a biparental burying beetle, *Nicrophorus orbicollis*. *Behaviour* 117 (1991): 82–105.

Insect Flight Mechanics and Beetle Flight

Dudley, R. *The Biomechanics of Insect Flight*. Princeton, N.J.: Princeton University Press, 2000.

Schneider, P. Die Flugtypen der Käfer (Coleoptera). *Entomologica Germanica* 1, nos. 3/4 (1975): 222–231.

Coloration and Mimicry

Anderson, T., and A. J. Richards. An electron microscope study of the structural colors of insects. *Journal of Applied Physiology* 13 (1942): 748–758.

Bagnara, J. *Chromatophores and Color Change*. Upper Saddle River, N.J.: Prentice-Hall, 1973.

Brower, L. P., J.V.Z. Brower, and P. W. Wescott. Experimental studies of mimicry, V: The reactions of toads (*Bufo terrestris*) to bumblebees (*Bombus americanum*) and their robberfly mimics (*Mallophora bomboides*) with a discussion of aggressive mimicry. *American Naturalist* 94 (1960): 343–355.

Cott, E. *Adaptive Colouration in Animals*. London: Methuen, 1940.

Evans, D. L., and G. P. Waldbauer. Behavior of adult and naïve birds when presented with a bumblebee and its mimics. *Zeitschrift für Tierpsychologie* 59 (1982): 247–259.

Fisher, R. M., and R. D. Tuckerman. Mimicry of bumble bees and cuckoo bees by carrion beetles (Coleoptera: Silphidae). *Journal of the Kansas Entomological Society* 59 (1986): 20–25.

Heinrich, B. A heretofore unreported color change in a beetle, *Nicrophorus tomentosus* Weber (Coleoptera: Silphidae). *Northeastern Naturalist* 19 (2012): 345–352.

Hinton, H. E., and G. M. Jarman. Physiological color change in the Hercules beetle. *Nature* 238 (1972): 160–161.

Lane, C., and M. A. Rothschild. A case of Muellerian mimicry of sound. *Proceedings of the Royal Entomological Society London A* 40 (1965): 156–158.

Prum, R. O., T. Quinn, and R. H. Torres. Anatomically diverse butterfly scales all produce structural colors by coherent scattering. *Journal of Experimental Biology* 209 (2006): 748–765.

Ruxton, G. D., T. N. Sherrett, and M. P. Speed. *Avoiding Attack: The Evolutionary Ecology of Crypsis, Warning Signals, and Mimicry*. New York: Oxford University Press, 2005.

Wickler, W. *Mimicry in Plants and Animals*. New York: McGraw-Hill, 1968.

Bumblebee Color Patterns

Heinrich, B. *Bumblebee Economics*. Cambridge: Harvard University Press, 1979; rev. ed., 2004.

Marshall, S. A. *Insects: Their Natural History and Diversity*. Buffalo,

N.Y.: Firefly Books, 2006. On insects in general, I particularly recommend this book.

Plowright, R. C., and R. E. Owen. The evolutionary significance of bumblebee color patterns: a mimetic interpretation. *Evolution* 34 (1980): 622–637.

Sendoff for a Deer

Forensic Entomology

Byrd, J. H., and J. L. Castner. *Forensic Entomology: The Utility of Arthropods in Legal Investigation.* Boca Raton, Fla.: CRC Press, 2001.

Dekeirsschieter, J., et al. Carrion beetles visiting pig carcasses during early spring in urban, forest and agricultural biotopes of Western Europe. *Journal of Insect Science* 11, no. 73 (2011).

The Ultimate Recycler: Remaking the World

Africa

Akeley, Carl. *In Brightest Africa.* Garden City, N.Y.: Doubleday, Page, 1923.

Huxley, Elspeth. *The Mottled Lizard.* London: Chatto & Windus, 1982.

van der Post, Laurens. *The Lost World of the Kalahari.* Middlesex, England: Penguin, 1958.

Roosevelt, Theodore. *African Game Trails.* New York: Charles Scribner's Sons, 1910.

Thomas, Elizabeth Marshall. *The Old Way: A Story of the First People.* New York: Picador, 2006.

Elephants

Joubert, Derek, and Beverly Joubert. *Elephants of Savuti.* National Geographic film.

Leuthold, W. Recovery of woody vegetation in Tsavo National Park, Kenya, 1970–1994. *African Journal of Ecology* 34, no. 2 (2008): 101–112.

Power, R. J., and R.X.S. Camion. Lion predation on elephants in the Savuti, Chobe National Park, Botswana. *African Zoology* 44 (2009): 36–44.

Hunting

Digby, Bassett. *The Mammoth and Mammoth Hunting in Northeast Siberia.* New York: Appleton, 1926.

Heinrich, B. *Why We Run: A Natural History.* New York: HarperCollins, 2001.

Jablonski, N. G. The naked truth. *Scientific American,* Feb. 2010: 42–49.

Lieberman, Daniel E., and Dennis M. Bramble. The evolution of marathon running: capabilities in humans. *Sports Medicine* 37 (2007): 288–290.

Peterson, Roger T., and James Fisher. *Wild America.* Boston: Houghton Mifflin, 1955.

Potts, Richard. *Early Hominid Activities at Olduvai.* New Brunswick, N.J.: Transaction Publishers, 1988.

Stanford, Craig B. *The Hunting Apes: Meat Eating and the Origins of Human Behavior.* Princeton, N.J.: Princeton University Press, 1999.

Predation

Darwin, Charles. "Diary of the Voyage of the H.M.S. *Beagle.*" In *The Life and Letters of Charles Darwin,* ed. Francis Darwin. London: D. Appleton, 1887.

Schaller, George B. *Serengeti Lion: A Study of Predator-Prey Relations.* Chicago: University of Chicago Press, 1972.

Schaller, George G. and Gordon R. Lowther. The relevance of carnivore behavior to the study of early hominids. *Southwestern Journal of Anthropology* 25 (1969): 307–41.

Schüle, Wilhelm. Mammals, vegetation and the initial human settlement of the Mediterranean islands: a palaeological approach. *Journal of Biogeography* 20 (1993): 399–412.

Stolzenberg, William. *Where the Wild Things Were: Life, Death, and Ecological Wreckage in a Land of Vanishing Predators.* New York: Bloomsbury, 2008.

Strum, Shirley C. Processes and products of change: baboon predatory behavior at Gilgil, Kenya. In *Omnivorous Primates*, ed. R. S. O. Harding and G. Teleki. New York: Columbia University Press, 1981.

Weapons

Guthrie, R. Dale. *The Nature of Paleolithic Art.* Chicago: University of Chicago Press, 2005.

Lepre, C. J., et al. An earlier origin for the Acheulian. *Nature* 477 (2011): 82–85.

Thieme, Hartmund. Lower Paleolithic hunting spears in Germany. *Nature* 385 (1997): 807–810.

The Overkill Hypothesis

Edmeades, Baz. *Megafauna — First Victims of the Human-Caused Extinctions* (www.megafauna.com, 2011). See chapter 13 for the debate about human scavenging and hunting, including the hunting of elephants.

Fiedel, Stuart, and Gary Haynes. A premature burial: comments on Grayson and Meltzer's "Requiem for overkill." *Journal of Archaeological Science* 31 (2004): 121–131.

Martin, P. S. Prehistoric overkill. In *Pleistocene Extinctions: The*

Search for a Cause, ed. P. S. Martin and H. E. Wright. New Haven: Yale University Press, 1967.

——. Prehistoric overkill: a global model. In *Quaternary Extinctions: A Prehistoric Revolution,* ed. P. S. Martin and R. G. Klein. Tucson: University of Arizona Press, 1989, pp. 354–404.

Surovell, T. A., N. M. Waguespack, and P. J. Brantingham. Global evidence for proboscidean overkill. *Proceedings of the National Academy of Sciences* 102 (2005): 6231–6336.

Northern Winter: For the Birds

Raven Reviews

Boarman, B., and B. Heinrich. Common raven (*Corvus corax*). In *Birds of North America,* no. 476, ed. A. Poole and F. Gill, pp. 1–32. Philadelphia: Academy of Natural Sciences, 1999.

Heinrich, B. Sociobiology of ravens: conflict and cooperation. *Sitzungberichte der Gesellschaft Naturforschender Freunde zu Berlin* 37 (1999): 13–22.

——. Conflict, cooperation and cognition in the common raven. *Advances in the Study of Behavior* 42 (2011).

Raven Carcass Scavenging

Heinrich, B. Dominance and weight-changes in the common raven, *Corvus corax. Animal Behaviour* 48 (1994): 1463–1465.

——. Winter foraging at carcasses by three sympatric corvids, with emphasis on recruitment by the raven, *Corvus corax. Behavioral Ecology and Sociobiology* 23 (1988): 141–156.

Heinrich, B., et al. Dispersal and association among a "flock" of common ravens, *Corvus corax. The Condor* 96 (1994): 545–551.

Heinrich, B., J. Marzluff, and W. Adams. Fear and food recognition in naive common ravens. *The Auk* 112, no. 2 (1996): 499–503.

Heinrich, B., and J. Pepper. Influence of competitions on caching behavior in the common raven, *Corvus corax*. *Animal Behaviour* 56 (1998): 1083–1090.

Marzluff, J. M., and B. Heinrich. Foraging by common ravens in the presence and absence of territory holders: an experimental analysis of social foraging. *Animal Behaviour* 42 (1991): 755–770.

Marzluff, J. M., B. Heinrich, and C. S. Marzluff. Roosts are mobile information centers. *Animal Behaviour* 51 (1996): 89–103.

Raven Intelligence, Cognition, and Communication

Bugnyar, T., and B. Heinrich. Hiding in food-caching ravens, *Corvus corax*. *Review of Ethology*, Suppl. 5 (2003): 57.

——. Food-storing ravens, *Corvus corax*, differentiate between knowledgeable and ignorant competitors. *Proceedings of the Royal Society London B* 272 (2005): 1641–1646.

——. Pilfering ravens, *Corvus corax*, adjust their behaviour to social context and identity of competitors. *Animal Cognition* 9 (2006): 369–376.

Bugnyar, T., M. Stoewe, and B. Heinrich. Ravens, *Corvus corax*, follow gaze direction of humans around obstacles. *Proceedings of the Royal Society London B* 271 (2004): 1331–1336.

——. The ontogeny of caching behaviour in ravens, *Corvus corax*. *Animal Behaviour* 74 (2007): 757–767.

Heinrich, B. Does the early bird get (and show) the meat? *The Auk* 111 (1994): 764–769.

——. Neophilia and exploration in juvenile common ravens, *Corvus corax*. *Animal Behaviour* 50 (1995): 695–704.

——. An experimental investigation of insight in common ravens, *Corvus corax*. *The Auk* 112 (1995): 994–1003.

——. Planning to facilitate caching: possible suet cutting by a common raven. *Wilson Bulletin* 111 (1999): 276–278.

Heinrich, B., and T. Bugnyar. Testing problem solving in ravens: string-pulling to reach food. *Ethology* 111 (2005): 962–976.

——. Just how smart are ravens? *Scientific American* 296, no. 4 (2007): 64–71.

Heinrich, B., and J. M. Marzluff. Do common ravens yell because they want to attract others? *Behavioral Ecology and Sociobiology* 28 (1991): 13–21.

Heinrich, B., J. M. Marzluff, and C. S. Marzluff. Ravens are attracted to the appeasement calls of discoverers when they are attacked at defended food. *The Auk* 110 (1993): 247–254.

Parker, P. G., et al. Do common ravens share food bonanzas with kin? DNA fingerprinting evidence. *Animal Behaviour* 48 (1994): 1085–1093.

Ravens and Wolves

Stahler, D. R., B. Heinrich, and D. W. Smith. The raven's behavioral association with wolves. *Animal Behaviour* 64 (2002): 283–290.

The Vulture Crowd

Wilbur, S. R., and J. A. Jackson, eds. *Vulture Biology and Management*. Berkeley: University of California Press, 1983. This volume, with forty contributors, is the last word on vultures and is said to "embody what is known about these birds today."

The Environment and Vulture Toxins

Albert, C. A., et al. Anticoagulant rodenticides in three owl species from Western Canada. *Archives of Environmental Contamination and Toxicology* 58 (2010): 451–459.

Layton, L. Use of potentially harmful chemicals kept secret under law. *Washington Post*, Jan. 4, 2010.

Magdoff, F., and J. B. Foster. What every environmentalist needs to know about capitalism. *Monthly Review* 61, no. 10 (2010): 11–30.

Peterson, Roger T., and James Fisher. *Wild America*. Boston: Houghton Mifflin, 1955, p. 301.

Vulture Guilds

Houston, D. C. Competition for food between Neotropical vultures in forest. *Ibis* 130, no. 3 (1988): 402–414.

Kruuk, H. J. Competition for food between vultures in East Africa. *Ardea* 55 (1967): 171–193.

Lemon, W. C. Foraging behavior of a guild of Neotropical vultures. *Wilson Bulletin* 103, no. 4 (1991): 698–702.

Wallace, M. P., and S. A. Temple. Competitive interactions within and between species in a guild of avian scavengers. *The Auk* 104 (1987): 290–295.

Vulture Decline

Gilbert, M. G., et al. Vulture restaurants and their role in reducing Diclofenec exposure in Asian vultures. *Bird Conservation International* 17 (2007): 63–77.

Green, R. E., et al. Diclofenac poisoning as a cause of vulture population declines across the Indian subcontinent. *Journal of Applied Ecology* 41 (2004): 793–800.

Markandya, A., et al. Counting the cost of vulture decline — an appraisal of human health and other benefits of vultures in India. *Ecological Economics* 67, no. 2 (2008): 194–204.

Prakash, V., et al. Catastrophic collapse of Indian white-backed *Gyps bengalensis* and long-billed *Gyps indicus* vulture populations. *Biological Conservation* 19, no. 3 (2003): 381–390.

——. Recent changes in populations of resident *Gyps* vultures in

India. *Journal of the Bombay Natural History Society* 104, no. 2 (2007): 129–135.

Swan, G. E., et al. Toxicity of Diclofenac to *Gyps* vultures. *Biology Letters* 2, no.2 (2006): 279–282.

Trees of Life

Mushrooms

The variety of fungi is endless, and there are numerous excellent books and guides for their identification, usually illustrated in color and with photographs. Some of my favorites, which have thousands of photographs of forty-one families of mushrooms, are the following.

Laessoe, T., A. Del Conte, and G. Lincoff. *The Mushroom Book: How to Identify, Gather, and Cook Wild Mushrooms and Other Fungi.* New York: DK Publishing, 1996.

Phillips, R. *Mushrooms of North America.* Boston: Little, Brown, 1991.

Roberts, P., and S. Evans. *The Book of Fungi: A Life-Size Guide to Six Hundred Species from Around the World.* Chicago: University of Chicago Press, 2011.

Stamets, Paul. *Mycelium Running: How Mushrooms Can Save the World.* New York: Ten Speed Press, 2005.

Decay of Trees

Dreistadt, S. H., and J. K. Clark. *Pests of Landscape Trees and Shrubs: An Integrated Pest Management Guide,* 2nd ed. Davis, CA: University of California Agriculture and Natural Resources, 2004.

Hickman, G. W., and E. J. Perry. *Ten Common Wood Decay Fungi in Landscape Trees: Identification Handbook.* Sacramento: Western Chapter, ISA, 2003.

Parkin, E. A. The digestive enzymes of some wood-boring beetle larvae. *Journal of Experimental Biology* 17 (1940): 364–377.

Shortle, W. C., J. A. Menge, and E. B. Cowling. Interaction of bacteria, decay fungi, and live sapwood in discoloration and decay of trees. *Forest Pathology* 8 (1978): 293–300.

Cetoniine Flower Beetles

Peter, C. I., and S. D. Johnson. Pollination by flower chafer beetles in *Eulophia ensata* and *Eulophia welwitchie* (Orchidacea). *South African Journal of Botany* 75 (2009): 762–770.

Utilization of Dead Wood

Evans, Alexander M. *Ecology of Dead Wood in the Southeast* (www.forestguild.org/SEdeadwood.htm), 2011. This scientific review, funded by the Environmental Defense Fund, includes about 200 references.

Kalm, Peter. *The America of 1750: Peter Kalm's Travels in North America*, vol. 1. Trans. from Swedish, ed. Adolph B. Benson. New York: Dover, 1937.

Kilham, L. Reproductive behavior of yellow-bellied sapsuckers. I. Preferences for nesting in *Fomes*-infected aspens and nest hole interrelations with flying squirrels, raccoons, and other animals. *Wilson Bulletin* 83, no. 2 (1971): 159–171.

Schmidt, M. M. I. et al. Persistence of soil organic matter as an ecosystem property. *Nature* 478 (2011): 49–56.

Dung Eaters

Bartholomew, G. A., and B. Heinrich. Endothermy in African dung beetles during flight, ball making, and ball rolling. *Journal of Experimental Biology* 73 (1978): 65–83.

Edwards, P. B., and H. H. Aschenbourn. Maternal care of a single off-spring in the dung beetle *Kheper nigroaeneus*: consequences of extreme parental investment. *Journal of Natural History* 23 (1975): 17–27.

Hanski, Ilkka, and Yves Cambefort, eds. *Dung Beetle Ecology*. Princeton, N.J.: Princeton University Press, 1990. An overview and review of dung beetle biology by multiple authors in relation to worldwide distribution, taxonomy, ecology, and natural history.

Heinrich, B., and G. A. Bartholomew. The ecology of the African dung beetle. *Scientific American* 241, no. 5 (1979): 146–156.

———. Roles of endothermy and size in inter- and intraspecific competition for elephant dung in an African dung beetle, *Scarabaeus laevistriatus*. *Physiological Zoology* 52 (1978): 484–494.

Ybarrondo, B. A., and B. Heinrich. Thermoregulation and response to competition in the African dung ball-rolling beetle *Kheper nigroaeneus* (Coleoptera: Scarabaeidae). *Physiological Zoology* 69 (1996): 35–48.

Elephants as Seed Dispersers

Campos-Arceiz, A., and S. Black. Megagardeners of the forest — the role of elephants in seed dispersal. *Acta Oecologica* (in press).

Ancient Scavenging Beetles

Chin, Karen, and B. D. Gill. Dinosaurs, dung beetles, and conifers: participants in a Cretaceous food web. *Palaios* 11, no. 3 (1996): 280–285.

Duringer, P., et al. First discovery of fossil brood balls and nests in the Chadian Pliovene Australopithecine levels. *Lethaia* 33 (2000): 277–284.

Grimaldi, D., and M. S. Engel. *Evolution of the Insects*. Cambridge, UK: Cambridge University Press, 2005.

Kirkland J. I., and K. Bader. Insect trace fossils associated with *Protoceratops* carcasses in the Djadokhta Formation (Upper Cretaceous), Mongolia. In *New Perspectives on Horned Dinosaurs: The Royal Tyrell Museum Ceratopsian Symposium*, ed. M. J. Ryan, B. J. Chinnery-Allgeier, and D. A. Eberth, pp. 509–519. Bloomington: Indiana University Press, 2010.

Beetles and Biocontrol

Bornemissza, G. F. An analysis of arthropod succession in carrion and the effect of its decomposition on the soil fauna. *Australian Journal of Zoology* 5 (1957): 1–12.

Michaels, K., and G. F. Bornemissza. Effects of clearfell harvesting on lucanid beetles (Coleoptera: Lucanidae) in wet and dry sclerophyll forests in Tasmania. *Journal of Insect Conservation* 3 (1999): 85–95.

Queensland Dung Beetle Project. Improving sustainable management systems in Queensland using beetles: final report of the 2001/2002 Queensland Dung Beetle Project (2002).

Sanchez, M. V., and J. F. Genise. Cleptoparasitism and detritivory in dung beetle fossil brood ball from Patagonia, Argentina. *Paleontology* 52 (2009): 837–848.

Salmon Death-into-Life

Salmon and Cycling

Hill, A. C., J. A. Stanford, and P. R. Leavitt. Recent sedimentary legacy of sockeye salmon (*Oncorhynchus nerka*) and climate change in an ultraoligotrophic, glacially turbid British Columbia nursery lake. *Canadian Journal of Fisheries and Aquatic Sciences* 66 (2009): 1141–1152.

Morris, M. R., and J. A. Stanford. Floodplain succession and soil nitrogen accumulation on a salmon river in southwestern Kamchatka. *Ecological Monographs* 81 (2011): 43–61.

Troll, Ray, and Amy Gulick. *Salmon in the Trees: Life in Alaska's Tongass Rain Forest.* Seattle: Braided River (Mountaineers Books), 2010.

Other Worlds

Chalk

Huxley, Leonard. *The Life and Letters of Thomas Henry Huxley.* New York: D. Appleton, 1901.

Huxley, T. H. On a piece of chalk. In *The Book of Naturalists,* ed. William Beebe. Princeton, N.J.: Princeton University Press, 1901.

Whale Falls

Little, Crispin T. S. The prolific afterlife of whales. *Scientific American* (Feb. 2010): 78–84.

Smith, Craig R., and Amy R. Baco. Ecology of whale falls at the deep-sea floor. In *Oceanography and Marine Biology: An Annual Review* 41 (2003): 311–354, ed. R. N. Gibson and R. J. A. Atkinson.

Thermal Vents

Cavanaugh, Colleen M., et al. Prokaryotic cells in the hydrothermal vent tube worm *Riftia pachyptila* Jones: possible chemoautotrophic symbionts. *Science* 213 (1981): 340–342.

Metamorphosis into a New Life and Lives

Metamorphosis

Ryan, Frank. *The Mystery of Metamorphosis: A Scientific Detective Story.* White River Junction, Vt.: Chelsea Green, 2011.

Truman, J. W., and L. M. Riddiford. The origin of insect metamorphosis. *Nature* 401 (1999): 447–452.

Wigglesworth, V. B. *The Physiology of Insect Metamorphosis.* Cambridge, UK: Cambridge University Press, 1954.

Williams, C. M. The juvenile hormone of insects. *Nature* 178 (1956): 212–213.

Hawkmoths

Kitching, I. J., and J. M. Cadiou. *Hawkmoths of the World.* Ithaca, N.Y.: Cornell University Press, 2000.

Larvae

Williamson, D. I. *The Origin of Larvae.* Boston: Kluwer Academic, 2003.

———. Hybridization in the evolution of animal form and life-cycle. *Zoological Journal of the Linnaean Society* 148 (2006): 585–602.

Beliefs, Burials, and Life Everlasting

Cambefort, Y. Le scarabée dans l'Egypte ancienne: origin et signification du symbole. *Révue de l'Histoire des Religions* 204 (1978): 3–46. Egyptian mummies inspired by dung beetles.

Robinson, A. How to behave beyond the grave. *Nature* 468 (2010): 632–633.

Schutz, E. Berichte über Geier als Aasfresser aus den 18. und 19. Jahrhundert. *Anzeiger der Ornithologischen Gesellschaft Bayern* 7 (1966): 736–738.

Schüz, Ernst, and Claus König. Old World vultures and man. In *Vulture Biology and Management,* ed. S. R. Sanford and A. L. Jackson. Berkeley: University of California Press, 1983, pp. 461–469.

Tibetan Customs

Hedin, S. *Transhimalaya*, vol. 1. Leipzig: Brockhaus, 1909.

Schafer, E. Ornithologische Forschungsergebnisse zweier Forschungs-
reisen nach Tibet. *Journal für Ornithologie* 86 (1938): 156–
166.

Taring, R. D. 1872. *Ich Bin Eine Tochter Tibets: Leben im Land der
vertriebenen Gotter*. Hamburg: Marion von Schröder, 1872.

Neolithic Vulture Cults

Lewis-Williams, D., and D. Pearce. *Inside the Neolithic Mind: Consciousness, Cosmos and the Realm of the Gods*, pp. 116–117. London: Thames and Hudson, 2003.

Mellaart, J. *Çatal Hüyük, a Neolithic Town in Anatolia*. London:
Thames and Hudson, 1967.

Mithen, Steven. *After the Ice: A Global Human History 20,000–5,000
BC*. Cambridge: Harvard University Press, 2004.

Selvamony, N. Sacred ancestors, sacred homes. In *Moral Ground:
Ethical Action for a Planet in Peril*, ed. K. D. Moore and M. P. Nel-
son. San Antonio, Tex.: Trinity University Press, 2010, pp. 137–140.

INDEX

Page references in italics refer to text illustrations.

use of old woodpecker holes, 123

oak: acorns, 40, 56; fungi in, 117; life span, 101; sapsuckers in, 121
octopus, 159, 180
odor of a carcass: attraction of scavengers by, 22; and burying beetle sense of smell, 31; chemical of putrefaction, 27; and fly sense of smell, 27, 31; stranded whale, 157
odors and scents: "calling" scent from burying beetle, 4, 12, 19; carcass (*see* odor of a carcass); shrew, as protection, 7; sphinx moth communication, 177
Oiceoptoma noveboracense, 32, *33*
oil and petroleum, 166, 168
ontogeny recapitulates phylogeny, 179–80
orchid, 115
Osedax (zombie worm), 163
Osmoderma scabra (scarab beetle), 113
otter, 154
owls: barn, 93; barred, 194; to control rodents, 93; food sources, 21, 93; great horned, 21; use of old woodpecker holes, 123
oxygen: in decomposition of tree, 125; lack in deepwater ocean, 161; recycling, xii; and respiration, 124; and the sphinx moth, 177–78

Pachylomera femoralis, 142–43
parasite, 33–34, 111, 146

parrot, 124
peat, 20, 168
Penicillium mold, 30
pesticides, 89, 90, 93
Peterson, Roger Tory, 27–28, 86–87
pets, 3, 83, 91
Phellinus igniarius (false tinder mushroom), 120–21
phosphorus, 128, 154
photosynthesis, 97, 163, 182–83. *See also* primary production
phylogeny, 179–80
physics and metaphysics, 190–94
pigeon, passenger, 20, 57
pigs: body temperature experiment, 28–29, 31; evolution, 49; food sources, 126; wild (boar), 41, 83
pines: bristlecone, 101; burrows and feeding tracks by beetles, 104–8, *106, 107*; to construct cabin, 103; white, 101, 103, 104–8, *106, 107*
pine sawyer beetles: boring into pine tree, 104–8, *106, 107*; description, 103, 104; eggs deposited on cut tree, 103; "sawing" sound, 105
plankton, 163–67, 180
plants: ecological role, xii–xiii, 97–99; pollinators, *114*, 114–15, 137, 175; seed dispersal, 137; *See also* trees
plastic, possible alternative, 146–47
"playing possum," 14, 15, 83
plesiosaur, 166
Pleurotus ostreatus (oyster mushroom), 117, *119*